Stochastic Filtering
with Applications in Finance

Stochastic Filtering
with Applications in Finance

Ramaprasad Bhar

The University of New South Wales, Australia

 **World Scientific**

NEW JERSEY · LONDON · SINGAPORE · BEIJING · SHANGHAI · HONG KONG · TAIPEI · CHENNAI

Published by

World Scientific Publishing Co. Pte. Ltd.

5 Toh Tuck Link, Singapore 596224

USA office: 27 Warren Street, Suite 401-402, Hackensack, NJ 07601

UK office: 57 Shelton Street, Covent Garden, London WC2H 9HE

Library of Congress Cataloging-in-Publication Data
Bhar, Ramaprasad.
  Stochastic filtering with applications in finance / by Ramaprasad Bhar.
  p. cm.
  Includes bibliographical references and index.
  ISBN-13: 978-981-4304-85-6
  ISBN-10: 981-4304-85-9
  1. Finance--Mathematical models. 2. Stochastic analysis. I. Title.
  HG106.B434 2010
  332.01'51922--dc22
                                        2010019089

British Library Cataloguing-in-Publication Data
A catalogue record for this book is available from the British Library.

Printed in Singapore by B & Jo Enterprise Pte Ltd

To Rajiv and Mitra

Preface

Stochastic filtering has been in use in engineering and in particular in the area of signal processing for a long time. Its usefulness in financial economics has been recognized for some time. The objective of a filtering application is to determine the best (in some sense) estimate for the states that drive a stochastic dynamical system. The measurable observations that the researchers have are not these states but some linear functions of these states. As an example, we can think of observations on futures prices on some asset, but we do not know anything about the stochastic convenience yield that may be contributing to the futures price formation. Filtering is an iterative process that enables us to estimate the states of the system as well as any hyper-parameters of the models. In the simplest case when there are no gaps in the measurable observations and all observations are made on equal time step, the solution to the problem is the well-known Kalman-Bucy filter. This basic structure, also referred to as state space model, has been found to be a valuable tool within Central Banks as well as financial institutions to deal with difficult inference problems as well for forecasting.

Complications arise when the functions relating to the unobservable states and the measurable observations are not linear. There may be missing observations or the conditional distributions are not necessarily Gaussian or even the time steps between the observations are not equal. To cater to such variations many improvements have been proposed in the literature.

The purpose of this book is to provide a comprehensive account of this very important tool with a view to making it more popular among researchers in finance and economics. It is not intended to give complete mathematical treatment of different stochastic filtering approaches, but describe them in simple terms and illustrate their application with real historical data for problems normally encountered in these disciplines. Although no prior knowledge in this area is assumed, the readers are expected to have knowledge of probability theory as well as general mathematical aptitude. The book is intended as a reference for graduate students and researchers interested in the field. It is also suitable for use as a text for a graduate level course on stochastic filtering.

The idea of such a book was growing since I started my own PhD research. I came to quantitative finance research after working in the information technology area for several years in technical software development and maintenance in commercial organizations. Besides, my early graduate level education in physics encourages a scientific frame of mind to deal with problems in financial economics. I attempted to synthesize the contents based on my life experiences in a diverse range of activities. I sincerely believe that this book will offer fresh perspectives to the readers in dealing with problems encountered in financial economics. The book is less demanding on mathematical aptitude and at the same time conveys the underlying complexities with the help of examples familiar to the readers in the finance and economics disciplines.

The book aims to be an example-driven elucidation of a complex technical area so that it appeals to a wide variety of readers in the discipline of finance and economics. The algorithmic presentation of the models for the applications will help readers to take up new challenges in their own field of study/research. This will also help them to get a clear understanding of the filtering issues quickly and enhance the models that suit their own perceptions.

The first chapter deciphers the notion of filtering and various ways of estimating models under different assumptions. Once this is understood, the succeeding chapters take up relevant applications in finance and economics and show how filtering framework applies in those situations. The second chapter focuses on the foreign exchange market and extracts embedded risk premium in the forward exchange rates. The third chapter explores the notion of risk premium in the equity market with a difference. It analyzes the price of risk and not just the total quantity of risk.

The fourth chapter revisits an important issue in macroeconomics. This is to do with the notion of inflation uncertainty and its connection with interest rates in the economy. It applies the filtering scheme to extract two forms of uncertainty from the inflation data and then explores its relationship with interest rates with particular reference to countries following inflation targeting policy.

The fifth and sixth chapters move the discussions to the interest rate model proposed by Heath, Jarrow and Morton in 1992. The rich but non-Markovian dynamics of this approach also entail computational and implementation complexities. These chapters describe how to deal with the issues via expansion of states and stochastic filtering. The examples include bond pricing as well as pricing of interest rate futures contracts.

Chapters 7 through 9 concentrate on fixed income securities. The possible drivers of credit spread are analyzed in Chapter 7 in a filtering framework. Chapter 8 takes up the structure of relatively new security in the credit derivatives market, i.e., credit default swaps. It disentangles the component structure of credit default swaps and analyzes the possible explanatory variables for these components. Chapter 9 moves on to explore the model for pricing options on credit default swaps. A relatively new filtering algorithm is exploited here for model calibration.

Chapter 10 explores different methods for estimation of the stochastic variance model applied to short-term interest rates. It also shows how to enhance the time series dynamics of the unobserved variance process.

Chapter 11 moves away from simply diffusion based filtering problems to those where jumps are also involved. This is illustrated with two examples. The first one is to capture the price path dynamics of spot electricity and the second one examines whether the jump risk could be diversified in the market for credit default swaps index and its sub-indices.

It is clear from the breadth of coverage in this book that it could not have been accomplished without the help from others. I would like to thank first my PhD students: Peipei Wang, Nedim Handzic and Damien Lee. I wish them all the success in their future endeavors. Also, I am grateful to a colleague of mine, David Colwell, with whom I had many insightful sessions.

I would also like to express my sincere gratitude to Carl Chiarella, my PhD supervisor at the University of Technology, Sydney. That is when I started learning about the interest rate model due to Heath, Jarrow and Morton and the complexities within the framework.

I am thankful to David Sharp at the World Scientific Press in Singapore for persisting with me to come up with the idea for this book.

Finally, I am grateful to my family for support and patience. My son regularly enquired about the progress of the book while he was studying Engineering at the Imperial College, London.

R. Bhar
Sydney, 2010

Contents

Preface   *vii*

1. Introduction: Stochastic Filtering in Finance
    1.1 Filtering Problem   2
    1.2 Examples of Filtering Applications   2
    1.3 Linear Kalman Filter   3
    1.4 Extended Kalman Filter (EKF)   6
    1.5 Applying EKF to Interest Rate Model   7
    1.6 Unscented Kalman Filter (UKF) for Nonlinear Models   10
    1.7 Background to Particle Filter for Non Gaussian Problems   13
    1.8 Particle Filter Algorithm   14
    1.9 Unobserved Component Models   16
    1.10 Concluding Remarks   19

2. Foreign Exchange Market – Filtering Applications
    2.1 Mean Reversion in Real Exchange Rates   21
    2.2 Common and Specific Components in Currency Movements   25
    2.3 Persistent in Real Interest Rate Differentials   30
    2.4 Risk Premia in Forward Exchange Rate   34
        2.4.1 Approach based on Market Price of Risk (BCP)   36
        2.4.2 Method of Wolff/Cheung   40
        2.4.3 Data and Empirical Results   41
        2.4.4 Summary of Section 2.4   43
    2.5 Concluding Remarks   47

3. Equity Market – Filtering Applications
    3.1 Introduction to Equity Price of Risk   49
        3.1.1 A Model for Equity Price of Risk   51
        3.1.2 Data Used for Empirical Study   52
        3.1.3 Discussion of Empirical Results   53
        3.1.4 Summary of Results   61

3.2 Economic Convergence in a Filtering Framework   62
  3.2.1 Defining Convergence   64
  3.2.2 Testing for Convergence   65
  3.2.3 Testing Convergence – Dickey-Fuller   66
  3.2.4 Testing Convergence – Kalman Filter   67
3.3 Ex-Ante Equity Risk Premium   69
  3.3.1 Background to Ex Ante Risk Premium   69
  3.3.2 A Model for Ex Ante Risk Premium   70
  3.3.3 Filtering Ex Ante Risk Premium   72
  3.3.4 Ex-Ante Risk Premium for UK   73
  3.3.5 Summarizing Ex-Ante Risk Premium for UK   73
3.4 Concluding Remarks   75

4. Filtering Application — Inflation and the Macroeconomy
  4.1 Background and Macroeconomic Issues   77
  4.2 Inflation Targeting Countries and Data Requirement   79
  4.3 Model for Inflation Uncertainties   80
  4.4 Testing Fisher Hypothesis   82
  4.5 Empirical Results and Analysis   83
  4.6 Concluding Remarks   85

5. Interest Rate Model and Non-Linear Filtering
  5.1 Background to HJM Model and the Related Literature   95
  5.2 The Basic HJM Structure   97
  5.3 Forward Rate Volatility: Deterministic Function of Time 100
  5.4 Forward Rate Volatility: Stochastic   102
  5.5 Estimation via Kalman Filtering   107
  5.6 Preference-Free Approach to Bond Pricing   109
  5.7 Concluding Remarks   112
    Appendix 5.1 Arbitrage-Free SDE for the Bond Price   114
    Appendix 5.2: Proof of Proposition 1   117
    Appendix 5.3: Proof of Proposition 2   119
    Appendix 5.4: Proof Proposition 3   122

6. Filtering and Hedging using Interest Rate Futures
  6.1 Background Details   126
  6.2 The Futures Price Model in the HJM Framework   127
  6.3 Non-Linear Filter for Futures Price System   131
  6.4 Data Used in Empirical Study   134
  6.5 Empirical Results   135
  6.6 Concluding Remarks   138
    Appendix 6.1   139

7. A Multifactor Model of Credit Spreads
   7.1 Background and Related Research 150
   7.2 Variables Influencing Changes in Credit Spreads 151
   7.3 Credit Spread and Default Risk 153
   7.4 Credit Spread and Liquidity 155
   7.5 Alternative Approach to Analyzing Credit Spread 156
   7.6 Data Used 159
   7.7 Multifactor Model for Credit Spread 160
   7.8 Fitting the Model 162
   7.9 Results 162
      7.9.1 Results for Apr-96 to Mar-03 162
      7.9.2 Results for Apr-96 to Mar-08 165
      7.9.3 Model Performance 168
      7.9.4 Discussion 168
   7.10 Concluding Remarks 169

8. Credit Default Swaps – Filtering the Components
   8.1 Background to Credit Default Swaps 185
   8.2 What is in the Literature Already? 188
   8.3 Credit Derivatives Market and iTraxx Indices 190
   8.4 CDS Index Data and Preliminary Analysis 192
   8.5 Focusing on Explanatory Variables 195
   8.6 Methodology for Component Structure 201
      8.6.1 Latent-Component Model for iTraxx Indices 201
      8.6.2 State Space Model and Stochastic Filtering 203
      8.6.3 Linear Regression Model for the Determinants of the CDS
      Components 204
   8.7 Analyzing Empirical Results 205
      8.7.1 Model Parameters and the Extracted Components 205
      8.7.2 Determinants of the Extracted Components 207
   8.8 Concluding Summary 211

9. CDS Options, Implied Volatility and Unscented Kalman Filter
   9.1 Background to Stochastic Volatility 230
   9.2 Heston Model in Brief 231
   9.3 State Space Framework 232
      9.3.1 Transition Equation 232
      9.3.2 Measurement Equation: CDS Option Price 233
      9.3.3 Measurement Equation Derivation 235
   9.4 General State Space Model and Filter Revisited 237
      9.4.1 Additive Non-Linear State Space model (Recap) 238
      9.4.2 The Scaled Unscented Transformation (Recap) 240

9.5 The Application of Unscented Kalman Filter   243
9.6 Empirical Results   245
9.7 Concluding Remarks   249

10. Stochastic Volatility Model and Non-Linear Filtering Application
10.1 Background to Stochastic Volatility Models   258
10.2 Stochastic Volatility Models of Short-term Interest Rates   259
10.2.1 SV-ARMA Specification   261
10.2.2 Exogenous Variables   262
10.3 Data for Analysis   263
10.4 Analysis of Estimation Results   264
10.5 Comparison of Volatility Estimates   266
10.6 Outline of State Space Model Estimation via MCL   271
10.7 Concluding Summary   273

11. Applications for Filtering with Jumps
11.1 Background to Electricity Market and Prices   285
11.2 A Model for Spot Electricity Prices   288
11.3 State Space Model, Kalman Filter and Poisson Jumps   291
11.4 Data and Empirical Results for Electricity Market   294
11.5 Summarizing Electricity Market Application   296
11.6 Background to Jumps in CDS Indices   297
11.7 CDS Data and Preliminary Analysis   300
11.8 Methodology for Analyzing CDS Jump Risks   301
11.8.1 Normality Test for CDS Index Distribution   301
11.8.2 Model for Individual iTraxx Indices   301
11.8.3 Multivariate Analysis of Jumps in iTraxx Index
with One Latent Common Factor   304
11.9 Analysis of Results from the CDS Market   307
11.10 Summarizing CDS Market Application   308

Bibliography   320

Index   337

Chapter 1

Introduction: Stochastic Filtering in Finance

This chapter introduces the background to the stochastic filtering problems as applied to various discipline area. This will point out different possible scenarios and solution approaches based on filtering for problems in finance and economics. This will also broadly outline the structures and models of a generalized stochastic filtering problem and the main equations that are needed. The introduction will also set the scene for various terminologies used in this context e.g. Unobserved Components (UC), Dynamic harmonic Regression (DHR), Forward Pass Kalman Filter, and Backward Pass Kalman Filter or Fixed Interval Smoothing (FIS). It will discuss further the extensions available for dealing with non-linear problems e.g. Extended Kalman Filter (EKF), and Unscented Kalman Filter (UKF).

Before ending introduction, it will give details of the contents of the following chapters and how these are organized etc. This relates to various application areas within the disciplines of finance and economics.

Some authors have identified three most successful areas of applications in quantitative finance. The first one relate to the unobserved volatility of any financial asset which is crucial to the development of derivatives products dependent on that asset. The second one concerns factor processes with or without jumps applied in commodities and interest rates. The third one concerns risk-minimizing hedging strategies under partial observations. Of course, there are several other areas of interests particularly in macro-economic issues.

1.1 Filtering Problem

The filtering problem deals with the estimation of a stochastic process X_t that is unobserved based on the past and current measurement of a related process Y, $\{Y_s : 0 \leq s \leq t\}$. The information coming out of the measurement process up to time t is conveniently represented by the σ-algebra, Υ_t generated by $\{Y_s : 0 \leq s \leq t\}$. Davis and Marcus (1981) give excellent introduction to the filtering problem. Besides, for a more complete and mathematical treatments see Lipster and Shiryayev (1978). On the other hand, Jazwinski (1970) gives a more applied and practical analyses of the filtering issues.

The filtering problem gives the solution to the question of conditional density of the signal X_t, $p(x_t | y_t)$ given the filtration Υ_t. In this context, Chaleyat-Maurel and Michel (1984) prove that the general solution is infinite dimensional. This implies that a general filter cannot be implemented by an algorithm that only uses a finite number of characteristics. In other words, there is no finite memory computer implementation that can be adequate for the filter. However, the most referenced implementation of the filter algorithm, Kalman filter, is a special case for the linear system with Gaussian conditional distribution. In this case, only two parameters mean and the variance, capture the essential attributes of $p(x_t | y_t)$.

1.2 Examples of Filtering Applications

Filtering has long been used in Control Engineering and Signal Processing. For the linear case, the Kalman filter is fast and easy to apply despite the noisiness and the length of the input data. It is an iterative procedure and the model parameters may depend upon a large number of observable and unobservable data.

In finance and economics applications we normally have a temporal time series of observations (Y_t). For example, Wells (1996) uses stock price series, Babbs and Nowman (1999) as well as Pennacchi (1991) use interest rate series, Lautier (2000) and Lautier and Galli (2000) use futures price series. The unobserved time series (X_t) in these cases could be volatility, convenience yield, price of risk etc depending on the objectives of the researchers. The iterative process allows us to link two consecutive unobservable states using a transition equation and a measurement equation relating to the hidden state. The idea is to proceed in two steps: first we estimate the hidden state, prior distribution, by using all the information up to that time-step. Then using this predicted value together with the new observation, we obtain a conditional estimation of the posterior distribution of the state.

It is useful at this stage to get used to some of the terminologies applicable to different filtering processes depending on the model characteristics. The linear filter, Kalman filters (KF), is applicable to cases when the transition and measurement equations are linear and the noise processes are Gaussian. With the same noise processes if the model equations are non-linear then we refer to this as the Extended Kalman Filter (EKF). Another more recent advance in this scenario is termed as the Unscented Kalman Filter (UKF). To deal with non-Gaussian cases we resort to Particle Filter (PF).

1.3 Linear Kalman Filter

In this section we describe the linear Gaussian filter and the algorithm that is needed for implementation. To assist with the exposition we will use a general specification of the model and will subsequently move on to more traditional specification more suitable for empirical studies.

Let the dynamic process X_t follows a transition equation

$$x_t = f\left(x_{t-1}, w_t\right), \tag{1.1}$$

and we also assume that we have a measurement Y_t such that

$$y_t = h\left(x_t, u_t\right). \tag{1.2}$$

In the above equations w_t and u_t are two mutually uncorrelated sequences of temporally uncorrelated sequences of Normal random variables with zero mean and covariance matrices Q_t and R_t respectively. Additionally, w_t is uncorrelated with x_{t-1} and u_t is uncorrelated with x_t.

We define the prior process estimate as

$$x_{t|t-1} = E[x_t] \tag{1.3}$$

which is the estimate of x_t at time t-1 just prior to making the measurement at time t. Similarly, we define the posterior estimate as

$$x_{t|t} = E[x_t \mid y_t] \tag{1.4}$$

which is the estimate at time t after the measurement at t has taken place. We also have the corresponding estimation errors $e_{t|t-1} = x_t - x_{t|t-1}$ and $e_t = x_t - x_t \mid y_t$. These give us the estimate of the error covariances as

$$P_{t|t-1} = E\left[e_{t|t-1} \cdot e'_{t|t-1}\right], \; P_{t|t} = E\left[e_t \cdot e'_t\right]. \tag{1.5}$$

In order to compute the above mean and the covariance we need the corresponding conditional densities $p(x_t \mid y_{t-1})$ and $p(x_t \mid y_t)$. These are determined iteratively via transition and measurement updates. The basic idea is to define the probability density function corresponding to the hidden state x_t given all the measurements made up to that time i.e. $y_{1:t}$.

The transition step is based upon Chapman-Kolmogorov equation

$$p(x_t \mid y_{1:t-1}) = \int p(x_t \mid x_{t-1}, y_{1:t-1}) p(x_{t-1} \mid y_{1:t-1}) dx_{t-1}$$
$$= \int p(x_t \mid x_{t-1}) p(x_{t-1} \mid y_{1:t-1}) dx_{t-1} \tag{1.6}$$

following the Markov property. The measurement update step is based upon Bayes rule

$$p(x_t \mid y_{1:t}) = \frac{p(y_t \mid x_t) p(x_t \mid y_{1:t-1})}{p(y_t \mid y_{1:t-1})} \tag{1.7}$$

and $p(y_t \mid y_{1:t-1}) = \int p(y_t \mid x_t) p(x_t \mid y_{1:t-1}) dx_t$. Additional details of this may be found in Jazwinski (1970).

At this point it is instructive to specialize the transition and the measurement equations (1.1) and (1.2) for a linear system and state the updating equations in a form amenable for easier implementation.

Let us focus on a linear state space system with transition equation of the form

$$x_t = T_t x_{t-1} + c_t + w_t \tag{1.8}$$

and the measurement equation

$$y_t = Z_t x_t + d_t + u_t \tag{1.9}$$

where c_t and d_t are possible time dependent vectors of compatible dimensions. Similarly the matrices T and Z_t are of dimensions compatible with the length of the state vector x_t and the measurement vector y_t respectively.

The actual algorithm may be stated as follows

Initialize x_0 and P_0

Transition equations updates for mean and covariance

$$x_{t|t-1} = T_t x_{t-1} + c_t \tag{1.10}$$

$$P_{t|t-1} = T_t P_{t-1} T_t' + Q_t \tag{1.11}$$

We define innovation as

$$v_t = y_t - Z_t x_t - d_t \tag{1.12}$$

Measurement equation update

$$x_{t|t} = x_{t|t-1} + K_t v_t \tag{1.13}$$

$$P_{t|t} = (I - K_t Z_t) P_{t|t-1} \tag{1.14}$$

$$K_t = P_{t|t-1} Z_t' \left(Z_t P_{t|t-1} Z_t' + R_t \right)^{-1} \tag{1.15}$$

and I is the identity matrix.

The matrix K_t is the Kalman gain and corresponds to the mean of the conditional distribution of x_t after making the measurement of y_t. It is the quantity that would minimize the mean square error P_t within the class of linear estimators.

This interpretation may be broadened further. Let us consider x as a Normally distributed variable with mean μ_x variance σ_{xx} and z another Normally distributed random variable with mean μ_z and variance σ_{zz}. The covariance between them is $\sigma_{xz} = \sigma_{zx}$. The conditional distribution of x|z is also Normal with $\mu_{x|z} = \mu_x + K(z - \mu_z)$ with Kalman gain corresponding to $K = \sigma_{xz} \sigma_{zz}^{-1}$.

In empirical applications most models would be parameterized and such parameters need to be calibrated to the available time series data. This calibration can be carried out by maximum likelihood method for conditionally Gaussian models. Let us label the unknown parameters by Θ.

We need to maximize the joint probability density $\prod_1^T p(y_t \mid y_{1:t-1})$. Since the conditional density is Normal and taking logarithm of the precise form of the function and ignoring the constant term the target likelihood function to be maximized is

$$L(\Theta) = -\sum_1^T \ln\left(\left|\Sigma_t\right|\right) - \sum_1^T v_t' \Sigma_t^{-1} v_t \qquad (1.16)$$

and $\Sigma_t = Z_t P_{t|t-1} Z_t' + R_t$.

1.4 Extended Kalman Filter (EKF)

In this section we discuss how to apply filtering algorithms when the assumption of Gaussian distribution still holds but the state equation and the measurement equation may be non-linear. It is easier to relate this discussion by focusing on the model given by equations (1.1) and (1.2). In the previous section the linear version used the model definition given by the equations (1.8) and (1.9). This section has benefited from Javaheri, Lautier, and Galli (2005).

The actual form of non-linearity given by (1.1) and (1.2) is not important to develop the ideas of Extended Kalman Filter (EKF). The functions in these equations need to linearized and use the notions of conservation of Normal distribution property within the class of linear functions. We, thus, define the Jacobian matrices of the function $f(.)$ with respect to the state variables and the system noise element and we refer to these as A_t and W_t. Precisely, for every row i and column j,

$$A_{ij} = \frac{\partial f_i\left(x_{t-1|t-1}, 0\right)}{\partial x_j}, \quad W_{ij} = \frac{\partial f_i\left(x_{t-1|t-1}, 0\right)}{\partial w_j}. \qquad (1.17)$$

Similarly, the Jacobain matrices for the measurement equation may be given by:

$$H_{ij} = \frac{\partial h_i\left(x_{t|t-1}, 0\right)}{\partial x_j}, \quad U_{ij} = \frac{\partial h_i\left(x_{t|t-1}, 0\right)}{\partial u_j}. \qquad (1.18)$$

These Jacobian matrices could be related to the system matrices of the linear system given by (1.8) and (1.9).

The transition and the updating equation of the previous section may now be expressed via these Jacobian matrices.

Transition equations updates for mean and covariance

$$x_{t|t-1} = f\left(x_{t-1|t-1}, 0\right) \tag{1.19}$$

$$P_{t|t-1} = A_t P_{t-1} \Lambda_t' + W_t Q_{t-1} W_t' \tag{1.20}$$

We define innovation as

$$v_t = y_t - h\left(x_{t|t-1}, 0\right) \tag{1.21}$$

Measurement equation updates are

$$x_{t|t} = x_{t|t-1} + K_t v_t \tag{1.22}$$

$$P_{t|t} = \left(I - K_t H_t\right) P_{t|t-1} \tag{1.23}$$

$$K_t = P_{t|t-1} H_t' \left(H_t P_{t|t-1} H_t' + U_t R_t U_t'\right)^{-1} \tag{1.24}$$

where I is the identity matrix.

Once this system of recursive equations are in place, the unknown parameters of the model can be calibrated using likelihood function maximization as outlined in the previous section. It may, however, be more instructive to follow an example due to Théoret, Rostan and El Moussadek (2004).

1.5 Applying EKF to Interest Rate Model

This illustration is based on the interest rate model due to Fong and Vasicek (1992). This two factor short rate model introduces the stochastic variance specification of Ornstein-Uhlenbeck form to the original Vasicek short rate model. This is described by

$$dr_t = \kappa\left(\mu - r_t\right)dt + \sqrt{v_t}\,dW_{r,t}$$
$$dv_t = \lambda\left(\bar{v} - v_t\right)dt + \tau\sqrt{v_t}\,dW_{v,t} \tag{1.25}$$

where $E\left(dW_{r,t}, dW_{v,t}\right) = \rho$, i.e. the two noise processes $W_{r,t}$ and $W_{v,t}$ are two correlated Brownian motions associated with the short rate and the short rate volatility processes, respectively.

The model allows stationary mean reverting process for the short rat where the variance of the short rate is also a stationary stochastic process. Here μ is the unconditional average short rate, κ is the speed of mean reversion in the short rate. Similarly, $\bar{v}$ is the average unconditional variance and λ defines the degree of persistence in the variance. Finally, τ is the variance of the unobserved variance process over a short period of time. The hidden volatility requires filtering technique to be extracted from observations on short rates.

The model in (1.25) is in continuous time as well as it is non-linear. Therefore, before applying EKF we need to transform it to discrete and approximate linear form.

The first of the two equations in (1.25) may be written in the differential notation as

$$d\left(e^{-\kappa t}(r_t - \mu)\right) = e^{\kappa t}\sqrt{v_t}\,dW_{r,t} . \tag{1.26}$$

Integrating by parts and assuming the sampling period being Δ this leads to,

$$r_{t+\Delta} = \mu + e^{-\kappa\Delta}(r_t - \mu) + e^{-\kappa\Delta}\int_t^{t+\Delta} e^{\kappa(s-t)}\sqrt{v_s}\,dW_{r,s} . \tag{1.27}$$

Similarly, for the variance equation,

$$v_{t+\Delta} = \bar{v} + e^{-\lambda\Delta}(v_t - \bar{v}) + e^{-\lambda\Delta}\tau\int_t^{t+\Delta} e^{\lambda(s-t)}\sqrt{v_s}\,dW_{v,s} . \tag{1.28}$$

If we now define

$$\varepsilon_t(\Delta) = e^{-\kappa\Delta}\int_t^{t+\Delta} e^{\kappa(s-t)}\sqrt{v_s}\,dW_{r,s} \tag{1.29}$$

and

$$\eta_t(\Delta) = e^{-\lambda\Delta}\int_t^{t+\Delta} e^{\lambda(s-t)}\sqrt{v_s}\,dW_{v,s} \tag{1.30}$$

then the equations (1.27) and (1.28) can be recast as the discrete form of the Fong and Vasicek model. Then,

$$r_{t+\Delta} = \mu + e^{-\kappa\Delta}(r_t - \mu) + \varepsilon_t(\Delta) \tag{1.31}$$

and

$$v_{t+\Delta} = \overline{v} + e^{-\lambda\Delta}\left(v_t - \overline{v}\right) + \eta_t\left(\Delta\right).$$ (1.32)

Now we focus on (1.29) and (1.30) and approximate the innovations as,

$$\varepsilon_{t\Delta}\left(\Delta\right) = e^{-\kappa\Delta}\sqrt{v_{t\Delta}}\sqrt{\Delta}\varepsilon_t, \quad \eta_{t\Delta}\left(\Delta\right) = e^{-\lambda\Delta}\tau\sqrt{v_{t\Delta}}\sqrt{\Delta}\eta_t$$ (1.33)

and $\varepsilon_t\left(\Delta\right)$ and $\eta_t\left(\Delta\right)$ are independent standard Normal random variables.

If we define a transformed discrete observed variable as,

$$R_t = e^{\kappa\Delta}\left(r_{(t+1)\Delta} - \mu\right) - \left(r_{t\Delta} - \mu\right), \ t=0,1,2...$$ (1.34)

and denote $V_t = v_{t\Delta}$ then the discrete state space form of the Fong and Vasicek model results,

$$R_t = \sqrt{\Delta}\sqrt{V_t}\varepsilon_t, \ t = 0,1,2,...$$
$$V_t = e^{-\lambda\Delta}V_{t-1} + \left(1 - e^{-\lambda\Delta}\right)\overline{v} + e^{-\lambda\Delta}\tau\sqrt{\Delta}\sqrt{V_{t-1}}\eta_t, \ t = 1,2,....$$ (1.35)

The first of the two equations in (1.35) is the measurement equation, but it is not linear in state variable for Kalman filter to be applicable. With a suitable change of variable this measurement equation could be made linear as, with

$$y_t = \ln\left(\frac{R_t^2}{\Delta}\right)$$ (1.36)

the first of equation (1.35) becomes,

$$y_t = \ln V_t + \ln \varepsilon_t^2.$$ (1.37)

Thus, the non-linear form of the measurement equation (1.37) and the second of the equation (1.35) create the discrete linear state space version of the Fong and Vasicek model. Since the measurement equation error term in (1.37) is not strictly Normal, it may be approximated as such and the EKF is now applicable.

Additional details of this model and the empirical results may be found in Théoret, Rostan and El Moussadek (2004).

1.6 Unscented Kalman Filter (UKF) for Nonlinear Models

In this section we describe another more recent development in application of Kalman filter to nonlinear systems. This new approach is due to Julier and Uhlmann (1997) and the description in this section follows the article by Javaheri, Lautier, and Galli (2005).

In this approach Jacobians are not computed. Instead the actual nonlinear system is used to approximate the distributions of the state and the measurement variables with a Normal distribution by applying unscented transformation to it. This allows more accurate evaluation of the mean and the covariance compared to that of the EKF method. The UKF method requires augmentation of the state vector with the system and the measurement noises and this leads to expanded state space. This is not required if the system noise and the measurement noise are linearly additive in their respective equations. Let us define this augmented state dimension as $na = nx + nw + nu$, where na, nx, nw, nu represent the augmented state space, original state space, dimension of the state noise and the dimension of the measurement noise respectively.

The UKF algorithm may be summarized as follows:

First initialize the state vector, x_0 and the state covariance P_0

Second, we need to define two weights to be associated with the mean $\omega_i(m)$, and the covariance $\omega_i(c)$ of the state as,

$$\omega_0(m) = \frac{\lambda}{na + \lambda}, \ \omega_0(c) = \frac{\lambda}{na + \lambda} + \left(1 - \alpha^2 + \beta\right) \qquad (1.38)$$

and for all other elements,

$$\omega_i(m) = \omega_i(c) = \frac{1}{2(na + \lambda)}, \ i = 1, 2 \ldots 2na \qquad (1.39)$$

where the scaling parameters α, β, κ and $\lambda = \alpha^2(na + \kappa) - na$ may be tuned for the particular problem. With these setting in place, the algorithm proceeds for each t as:

Third, the augmented state vector is created by concatenating the original state vector, state noise and the measurement noise,

$$x_{t-1}^a = \begin{bmatrix} x_{t-1} \\ w_{t-1} \\ u_{t-1} \end{bmatrix}. \tag{1.40}$$

Thus the augmented state expectation after a measurement,

$$x_{t-1|t-1}^a = E\left[x_{t-1}^a \mid y_{k-1}\right] = \begin{bmatrix} x_{t-1|t-1} \\ 0 \\ 0 \end{bmatrix} \tag{1.41}$$

and similarly, the state covariance,

$$P_{t-1|t-1}^a = \begin{bmatrix} P_{t-1|t-1} & P_{xw(t-1|t-1)} & 0 \\ P_{xw(t-1|t-1)} & P_{ww(t-1|t-1)} & 0 \\ 0 & 0 & P_{uu(t-1|t-1)} \end{bmatrix} \tag{1.42}$$

where the subscripts x, w, and u correspond to the original state vector, state noise and the measurement noise parts respectively.

Fourth, to achieve the Normal approximation we need to define what is known as the Sigma Points for Unscented Transformation:

$$\chi_{t-1}^a(0) = x_{t-1|t-1}^a \tag{1.43}$$

and,

$$\chi_{t-1}^a(i) = x_{t-1|t-1}^a + \left((na+\lambda)P_{t-1|t-1}^a\right)_i^{0.5}, \; i = 1,...na, \tag{1.44}$$

$$\chi_{t-1}^a(i) = x_{t-1|t-1}^a - \left((na+\lambda)P_{t-1|t-1}^a\right)_{i-na}^{0.5}, \; i = na+1,...2na \tag{1.45}$$

where the subscripts i and (i - na) in equations (1.44) and (1.45) refer to the corresponding columns of the matrix in the square root.

Fifth, the augmented state equation updates may be stated as,

$$\chi_{t|t-1}(i) = f\left(\chi_{t-1}^x(i),\chi_{t-1}^w(i)\right), \; i = 0,...2na+1 \tag{1.46}$$

$$x_{t|t-1} = \sum_{i=0}^{2na} \omega_i(m)\chi_{t|t-1}(i) \tag{1.47}$$

$$P_{t|t-1} = \sum_{i=0}^{2na} \omega_i(c)\left(\chi_{t|t-1}(i) - x_{t|t-1}\right)\left(\chi_{t|t-1}(i) - x_{t|t-1}\right)' \tag{1.48}$$

and the superscript x and w in the above equations represent the state and the state noise part of the augmented state vector, respectively.

Sixth, the innovation from the measurement equation may be written as,

$$y_{t|t-1}(i) = h\left(\chi_{t|t-1}(i), \chi_{t-1}^{u}(i)\right) \tag{1.49}$$

$$y_{t|t-1} = \sum_{i=0}^{2na} \omega_i(m) y_{t|t-1}(i) \tag{1.50}$$

and,

$$v_t = y_t - y_{t|t-1}. \tag{1.51}$$

Seventh, the measurement updates are,

$$P_{y_t y_t} = \sum_{i=0}^{2na} \omega_i(c)\left(y_{t|t-1}(i) - y_{t|t-1}\right)\left(y_{t|t-1}(i) - y_{t|t-1}\right)', \tag{1.52}$$

$$P_{\chi_t y_t} = \sum_{i=0}^{2na} \omega_i(c)\left(\chi_{t|t-1}(i) - x_{t|t-1}\right)\left(y_{t|t-1}(i) - y_{t|t-1}\right)', \tag{1.53}$$

where Kalman gain is given by,

$$K_t = P_{\chi_t y_t} P_{y_t y_t}^{-1}. \tag{1.54}$$

Finally, as in the linear system,

$$x_{t|t} = x_{t|t-1} + K_t v_t \tag{1.55}$$

and,

$$P_{t|t} = P_{t|t-1} - K_t P_{y_t y_t} K_t'. \tag{1.56}$$

This completes all the recursions needed to develop the likelihood function that needs to be maximized in order to estimate the unknown parameters in the model.

In a later chapter we will examine the application of UKF to a practical problem.

1.7 Background to Particle Filter for Non Gaussian Problems

In this section we discuss the basic idea behind particle filer and the next section will develop the filtering algorithm utilizing this concept. Instead of using a Gaussian approximation to infer $(x_t \mid y_t)$ it employs Monte Carlo Simulation to achieve that. This requires deciding on a proposal distribution, $q(.)$, which is simple and importance sampling as described below.

In term of the proposal distribution the expected value will be given by,

$$E[f(x_t)] = \int f(x_t) p(x_t \mid y_{1:t}) dx_t$$
$$= \int f(x_t) \frac{p(x_t \mid y_{1:t})}{q(x_t \mid y_{1:t})} q(x_t \mid y_{1:t}) dx_t \,. \tag{1.57}$$

This may be re-expressed as,

$$E[f(x_t)] = \int f(x_t) \frac{w_t(x_t)}{p(x_t \mid y_{1:t})} q(x_t \mid y_{1:t}) dx_t,$$
$$w_t(x_t) = \frac{p(x_t \mid y_{1:t}) p(x_t)}{q(x_t \mid y_{1:t})} \,. \tag{1.58}$$

This is referred to as the non-normalized weight during filtering at step t.

We can convert this to normalized weight as,

$$E[f(x_t)] = \frac{E_q[w_t(x_t) f(x_t)]}{E_q[w_t(x_t)]} \equiv E_q[\tilde{w}_t(x_t) f(x_t)]$$
$$\tilde{w}_t(x_t) = \frac{w_t(x_t)}{E_q(x_t \mid y_t)} \,. \tag{1.59}$$

We can now apply Monte Carlo Simulation (with N sampling points) from the distribution of $q(x_t \mid y_{1:t})$ in discrete framework to compute the above expectation,

$$E[f(x_t)] \approx \sum_{i=1}^{N} \tilde{w}_t(x_t^i) f(x_t^i), \quad \tilde{w}_t(x_t^i) = \frac{w_t(x_t^i)}{\sum_{i-1}^{N}(x_t^i)} \,. \tag{1.60}$$

We now need a mechanism to recursively compute this weight from one time step to the next. This is achieved by assuming that the proposal distribution is Markov. This recursion is,

$$w_t^i = w_{t-1}^i \frac{p\left(y_t \mid x_t^i\right) p\left(x_t^i \mid x_{t-1}^i\right)}{q\left(x_t^i \mid x_{t-1}^i, y_{1:t}\right)} . \tag{1.61}$$

This now completely defines the recursion for the importance sampling procedure and it solely relies on filtering and not smoothing as may be applicable to Kalman filter.

We will conclude this section with some observations on practical issues related to implementation of this re-sampling procedure. The variance of the weights calculated above may vary widely and thus several suggestions have been proposed in the literature. Some of these could be found in Javaheri, Lautier, and Galli (2005). One such suggestion is to use Gaussian approximation for the proposal distribution based on the Kalman filter discussed earlier. This has the added advantage of utilizing the most current observation in the observed series. In term of earlier notation this may be specified as,

$$q\left(x_t \mid x_{t-1}, y_{1:t}\right) = N\left(x_{t|t}, P_{t|t}\right) . \tag{1.62}$$

Further information can be obtained from the cited references.

1.8 Particle Filter Algorithm

In this section we discuss the enhancement to the filtering algorithm needed to accommodate particle filter. This applies to both the EKF and the UKF described earlier.

At t=0, we first set x_0 and P_0 similar to earlier algorithms.
For each point of the N sampling points of the simulation we set,

$$x_0^i = x_0 + \sqrt{P_0}\, z^i \tag{1.63}$$

where z is a standard Normal simulated number. Also, let

$$P_0^i = P_0, \quad w_0^i = \frac{1}{N} .$$

For each of the simulation points (i), follow the steps:

$$x^i_{t|t} = KF\left(x^0_{t-1|t-1}\right)$$

and similarly the associated covariance $P^i_{t|t}$ computed via the EKF or the UKF. KF indicates this association with the EKF or the UKF.

For each simulation point again,

$$\tilde{x}^i_t = x^i_{t|t} + \sqrt{P^i_0}\, z^i$$

with z being a simulated standard Gaussian number.

Now, calculate the weight for each of the simulation points following,

$$w^i_t = w^i_{t-1}\, \frac{p\left(y_t \mid \tilde{x}^i_t\right) p\left(\tilde{x}^i_t \mid x^i_{t-1}\right)}{q\left(\tilde{x}^i_t \mid x^i_{t-1}, y_{1:t}\right)}$$

where $q(.)$ is the proposal Normal density with mean $x^i_{t|t}$ and covariance $P^i_{t|t}$.

We now need to normalize the weights as,

$$\tilde{w}^i_t = \frac{w^i_t}{\sum^N_{i=1} w^i_t}$$

and resample the points $\tilde{x}^i_t$, get x^i_t and reset the weights as

$$w^i_t = \tilde{w}^i_t = \frac{1}{N}$$

and repeat the steps for the next time step.

Using the particle filter we can also estimate the unknown model parameters by maximizing the likelihood function as before. Since, by assumption the distribution is not Gaussian in this case; we need some understanding how to formulate the problem.

The likelihood in step t is,

$$l_t = p\left(y_t \mid y_{1:t-1}\right) = \int p\left(y_t \mid x_t\right) p\left(x_t \mid y_{1:t-1}\right) dx_t \qquad (1.64)$$

the total likelihood is the product of the above for all t. However, (1.64) may also be written as,

$$l_t = \int p(y_t \mid x_t) \frac{p(x_t \mid y_{1:t-1})}{q(x_t \mid x_{t-1}, y_{1:t})} q(x_t \mid x_{t-1}, y_{1:t}) dx_t \qquad (1.65)$$

and since by construction $\tilde{x}_t^i$ are distributed as $q(.)$, and resetting $\tilde{w}_t^i$ to a constant (1/N) during re-sampling, it is possible to approximate l_t as,

$$\tilde{l}_t = \sum_{i=1}^{N} w_t^i . \qquad (1.66)$$

This gives an alternative interpretation the total likelihood.

1.9 Unobserved Component Models

This section focuses on adaptive, off-line signal processing and forecasting for non-stationary time series described by the following Unobserved Component model (e.g. Young et al, 1998):

$$y_t = T_t + S_t + g(u_t) + e_t \qquad (1.67)$$

where, y_t as before is the observed time series, T_t is the low frequency or trend component, S_t is a periodic component, possibly displaying temporal change in both amplitude and phase, $g(u_t)$ captures the influence of a set of exogenous variables and e_t is an irregular component assumed to be Normally distributed with mean zero and variance σ^2.

In financial economic applications most time series we encounter are non-stationary, or first difference stationary. In these situations, it is customary to define the trend component in (1.67) as a random walk. Cochrane (1988) uses the Beveridge and Nelson (1981) decomposition to express a first-difference stationary process as the sum of (covariance) stationary and random walk components. He argues that a measurement of the size of the random walk component can be a better guide to the proper statistical characterization of the series than a simple unit root test. He proposes a non-parametric method, the Variance Ratio, for determining the magnitudes of the random walk and stationary components of a time series. Further insight may be gained from Cochrane's article. Fama and French (1988) utilize similar concept for stock market application with non-stationary trend and stationary temporal components. Hatgioannides, Karanasos and Karanassou (2004) also apply the same intuition to term structure of interest rate model.

Here, we proceed to understand the issues relating to filtering in such a set up as in (1.67).

In (1.67) several components of the model are present and, in general, it may lead to identification problem. However, in the special case below the model remains identifiable. This is referred to as the Dynamic Harmonic Regression (DHR), Young (1988).

$$y_t = T_t + S_t + e_t, \; e_t \sim N\left(0, \sigma^2\right). \tag{1.68}$$

In (1.68) the most important unobserved component is the term S_t defined as,

$$S_t = \sum_{i=1}^{M}\left\{a_{i,t}\cos\left(\omega_{i,t}\right) + b_{i,t}\sin\left(\omega_{i,t}\right)\right\} \tag{1.69}$$

where $a_{i,t}, b_{i,t}$ are time varying parameters and ω_i represents fundamental and harmonic frequencies associated with the periodicity of the series. The frequency range could be chosen to match the spectral properties of the time series and discussed in detail in Young (1988). The DHR model is, therefore, time dependent extension of the classical Fourier series.

Each of the time varying components and the trend component (T_t) in (1.68) could be represented by a state vector of size 2, $x_i \equiv \left[l_i, d_i\right]'$, where the two components represent the changing level and the slope of the associated trend or time varying parameter. Usually, the stochastic evolution of each $x_i \equiv \left[l_i, d_i\right]'$ is assumed to be driven by a Generalized Random Walk (GRW) of the form,

$$x_{i,t} = \begin{bmatrix} \alpha & \beta \\ 0 & \gamma \end{bmatrix} x_{i,t-1} + \begin{bmatrix} \delta & 0 \\ 0 & 1 \end{bmatrix} \eta_{i,t}, \; i=1,\dots M+1. \tag{1.70}$$

With appropriate constraints on the elements of the matrices in (1.70) different Random Walk models could be realized. For example, Integrated Random Walk would result if, $\alpha = \beta = \gamma = 1, \delta = 0$. Scalar Random Walk would results if $\alpha = 1, \beta = \gamma = 0, \delta = 1$. Similarly, Local Linear Trend results if $\alpha = \beta = \gamma = \delta = 1$. Further discussions about this can be found in Young et al (1989).

The overall state space model can be described as an example, assuming that M=1. The main idea is to stack the system matrices in (1.70) such a way that the overall state equation is obtained. Similarly, by suitable choice of the measurement matrices the measurement equation of (1.68) is obtained. For M=1 this results in the state equation,

$$x_t \equiv \begin{bmatrix} l_{1,t} \\ d_{1,t} \\ l_{2,t} \\ d_{2,t} \end{bmatrix} = \begin{bmatrix} \alpha_1 & \beta_1 & 0 & 0 \\ 0 & \gamma_1 & 0 & 0 \\ 0 & 0 & \alpha_2 & \beta_2 \\ 0 & 0 & 0 & \gamma_2 \end{bmatrix} x_{t-1} + \begin{bmatrix} \delta_1 & 0 & 0 & 0 \\ 0 & 1 & 0 & 0 \\ 0 & 0 & \delta_2 & 0 \\ 0 & 0 & 0 & 1 \end{bmatrix} \begin{bmatrix} \eta_1 \\ \eta_1 \\ \eta_2 \\ \eta_2 \end{bmatrix} \quad (1.71)$$

and the measurement equation is,

$$y_t = \begin{bmatrix} 1 & 1 & 1 & 1 \end{bmatrix} x_t + e_t. \quad (1.72)$$

For the special case of M=1, the equations (1.71) and (1.72) are in the state space form and the algorithm described in an earlier section for Linear Kalman Filter can be applied to estimate the model parameters by maximizing the likelihood function.

To complete our discussions on filtering algorithms we outline here the Backward Pass Filter or Fixed Interval Smoothing algorithm. The algorithm described in section 1.3 in connection with Linear Kalman Filter is also referred to as Forward Pass Filter. For most applications in financial economics the time series of observations have already been made. In other words, the researcher is attempting to fit a model a historical time series. By applying the Forward Pass Filter the model parameters are obtained by maximizing the likelihood function. Also, at the same time, since the complete observation data set is available it is possible to refine the estimates via the Backward Pass Filter by running the procedure from the end of the time series to the beginning.

We define the recursive relations for the Backward Pass Filter with reference to the model in (1.8) and (1.9). Starting with the initial condition at the last measurement point, $y_{T|T}$ and $P_{T|T}$

$$y_{t-1|T} = y_{t-1|t-1} + J_{t-1}\left(y_{t|T} - y_{t|t-1}\right) \quad (1.73)$$

$$P_{t-1|T} = P_{t-1|t-1} + J_{t-1}\left(P_{t|T} - P_{t|t-1}\right)J'_{t-1} \quad (1.74)$$

where,

$$J_{t-1} = P_{t-1|t-1} T' \left[P_{t|t-1} \right]^{-1} . \qquad (1.75)$$

It should be clear from the above that to implement the Backward Pass Filter the quantities $y_{t|t}$ and $P_{t|t}$ generated during the Forward Pass Filter must be saved.

This completes our exposition on filtering algorithms and those interested in learning more about applications in engineering (signal processing) the references cited in this section are excellent source.

1.10 Concluding Remarks

We have provided an exposition of various filtering scheme and the main algorithm to implement these for a typical application in the discipline of financial economics. This includes Linear Kalman Filter and its extensions e.g. Extended Kalman Filter applicable to non-linear Gaussian models, Unscented Kalman Filter an alternative to EKF but with more precision as well as Particle Filter to deal with non-Gaussian models. At the end we have also introduced Dynamic Harmonic Regression to cover unobserved component models with time varying parameters and possibly non-stationary time series. The focus has mainly been in the algorithms and the conceptual background to these sophisticated approaches to stochastic modeling.

In the following chapters we are going to discuss various applications that utilize such algorithms. These applications will relate to foreign exchange markets, stock markets, interest rate models as wells as market for fixed income securities. These chapters will include more of economic intuitions in the estimated models – parameters as well as the filtered components. There would be sufficient background to the development of the models for different markets, but the filtering process would rely on the materials in chapter 1.

Chapter 2

Foreign Exchange Market – Filtering Applications

This chapter focuses on the foreign exchange market and explores how researchers have applied stochastic filtering methods to seek answers for various issues. The emphasis of the presentation would be the modeling approach to problem being examined rather than analyzing empirical results obtained by the authors. We will, however, outline the main points of the empirical results.

There are four main topics examined here. First, the concern is the behavior of the mean reversion pattern in real exchange rates in countries that have experienced high inflation episodes in the past. Second, we look at common and country specific movements in the exchange rates of a similar set of countries in the same geographical region. Third, an aspect of real interest rate differential is examined. This also highlights the existence of unobserved risk premium in the forward exchange rates. Fourth, we illustrate the approaches to infer such unobserved risk premium from available exchange rates.

2.1 Mean Reversion in Real Exchange Rates

This topic is based upon the article by Bleaney, Leybourne and Mizen (1999). The question of differences in exchange rate dynamics during periods of high inflation has not been sufficiently explored. Yet there is a strong prima facie case for the proposition that the dynamics of real exchange rates are influenced by inflationary circumstances. There could two reasons for this. First, at high rates of inflation, nominal shocks dominate real shocks, whereas at low rates of inflation the opposite is likely to be true. This suggests that mean-reverting tendencies in real exchange rates are expected to be more evident at higher inflation regimes. Second, when inflation is high, demand for domestic money as an asset falls and it begins to be displaced by foreign currencies. This currency substitution implies flows across the exchanges, which are likely to affect both the short-run dynamics and the long-run equilibrium in real exchange rates.

Several empirical literatures show that the purchasing power parity (PPP) appears to hold during episodes of high inflation. Taylor and McMahon (1988) document this for the German hyperinflationary experience of 1922-1923. McNown and Wallace (1989) and Liu (1992) document the same phenomenon for pegged exchange rates in Latin American countries in the postwar period. In those studies the estimated coefficients of the co-integrating regressions between exchange rates and relative prices are much closer to the PPP-predicted value of unity in high inflation cases.

Empirical evidence from countries that have experienced varying inflation rates is, however, more ambiguous. Zhou (1997), for example, cannot reject a unit root in the real exchange rate for five countries with episodes of high inflation. When the structural breaks are accounted for, Zhou concludes in favor of stationarity. These structural breaks represent shifts in the level and/or the time trend of the estimated equilibrium real exchange rate. These shifts appear to be associated with changes in the inflation regime.

In these empirical studies most researchers employ fixed-parameter mean reversion models of the real exchange rate using the standard augmented Dickey-Fuller (ADF) unit root tests and compare these with the Leybourne-McCabe test, which takes stationarity as the null and has a unit root alternative. It is also possible to allow for a stochastic unit root and permit the deviations of the root from unity to follow a noise process or a random walk. This is where filtering becomes an unavoidable tool to investigate the unobserved trajectory of the root over time.

Using monthly exchange rate data and wholesale price index, Bleaney, Leybourne and Mizen (1999) find possible evidence of both kinds of behavior. They are able to show that large spikes occur in the mean reversion process at times of high inflation. The countries examined by these authors are Argentina, Chile, Colombia and Israel.

Models of Mean Reversion:

Bleaney, Leybourne and Mizen (1999) take the nominal end-of-month exchange rate for the high inflation countries vis-à-vis the U.S. dollar $\left(e_t\right)$, the domestic price level $\left(p_t\right)$, and the U.S. price level $\left(p_t^*\right)$, where all variables are in natural logarithms. The real exchange rate can be defined as

$$s_t = e_t - p_t + p_t^* . \tag{2.1}$$

Bleaney, Leybourne and Mizen (1999) first consider the time series properties of s_t employing both unit root tests and stationarity tests. They then test for the presence of stochastic unit roots in the data. Series that appear to contain stochastic unit roots by casting the model in state space form and by applying Linear Kalman Filter algorithm. This allows them to examine the trajectory of the root through time.

A simple linear model for unit roots and stationarity may be stated as,

$$s_t = (1 + \delta) s_{t-1} + \varepsilon_t \tag{2.2}$$

where, ε_t is a stationary process, allows them to test the hypothesis whether the real exchange rate process contains a unit root when $\delta = 0$. On the other hand, if $\delta < 0$ then the real exchange rate series is stationary with a constant rate of mean reversion. These two tests could be contrasted by using ADF test.

Alternatively, another model,

$$s_t = w_t + \varepsilon_t, \ \Delta w_t = \eta_t, \ \eta_t \sim \text{i.i.d.}\left(0, \sigma_\eta^2\right) \tag{2.3}$$

may be used. If $\sigma_\eta^2 = 0$ then the process is stationary with a constant rate of mean reversion, and for $\sigma_\eta^2 > 0$ the re is zero rate of mean reversion.

If these above tests show that there is non-stationarity in the data, then the following model could be a useful description of the data generating process:

$$s_t = \left(1 + \delta_t\right)s_{t-1} + \varepsilon_t, \ \delta_t \sim \text{i.i.d.}\left(0, \sigma_\delta^2\right). \tag{2.4}$$

The above equation is said to contain a stochastic unit root. Granger and Swanson (1997) show that this model accommodates stochastic unit root. The process can be thought of as having mean reversion in some periods and mildly explosive behavior in other periods. That is a reasonable characterization of real exchange rates. This allows the test for a fixed unit root $\left(\sigma_\delta^2 = 0\right)$ against a stochastic unit root $\left(\sigma_\delta^2 > 0\right)$. For the fixed unit root case it is difference stationary, but non-stationary under the alternative.

Leybourne, McCabe, and Mills (1996) suggest an alternative to (2.4) using a random walk specification for δ_t,

$$s_t = \left(1 + \delta_t\right)s_{t-1} + \varepsilon_t, \ \Delta \delta_t = \mu_t, \ \mu_t \sim \text{i.i.d.}\left(0, \sigma_\mu^2\right). \tag{2.5}$$

This specification allows persistence in deviation from unit root. In reality, the actual process generating deviations from unit root may lie somewhere between the two specifications (2.4) and (2.5). To allow for this we may need to allow δ_t following an unrestricted autoregressive process.

As pointed out by Granger and Swanson (1997), the standard ADF test for unit root will tend to indicate its presence when the process actually is a stochastic unit root rather than stationary. Similar problem may exist for unit root testing via (2.4) and (2.5). Considering all these points, Bleaney, Leybourne and Mizen (1999), suggest the following specification in the state space framework,

$$\left(s_t - \alpha - \sum_{i=1}^{l} \phi_i s_{t-i}\right) = \left(1 + \delta_t\right)\left(s_{t-1} - \alpha - \sum_{i=1}^{l} \phi_i s_{t-i-1}\right) + v_t \tag{2.6}$$

$$\delta_t = \rho \delta_{t-1} + \omega_t \tag{2.7}$$

$$v_t \sim \text{i.i.d.}\left(0, \sigma_v^2\right),\ \omega_t \sim \text{i.i.d.}\left(0, \sigma_\omega^2\right). \tag{2.8}$$

In this above specification the stochastic deviations from the unit root, δ_t, follow an unrestricted first-order autoregressive process. A fixed unit root process is a special case of this model and arises if $\sigma_\omega^2 = 0$. If we assume that v_t and ω_t are independent and normally distributed, then the model of (2.6) is conditionally Gaussian and its likelihood function can be constructed via the filter algorithm of Linear Kalman Filter described in chapter 1.

Bleaney, Leybourne and Mizen (1999) apply this technique to the four countries using monthly data spanning 1972 to 1993. The estimation results are given in Table 2.1 and they also compute the smoothed values of the state variable, δ_t. Here, we reproduce their table and include a brief analysis of the results.

The Table 2.1 gives the parameter estimates together with the t-statistics in parentheses below the parameters. One point is clear from this table that it rules out fixed mean reversion since $\sigma_\omega^2 > 0$ for all the four countries and are statistically significant. The autoregressive coefficients of the unobserved δ_t are also significant for all the countries. For more information on the smoothed estimates of the state variable δ_t readers are encouraged to see Bleaney, Leybourne and Mizen (1999).

Table 2.1 Parameter Estimates of Equation 2.6.

	Argentina	Chile	Colombia	Israel
σ_ω^2	0.043	0.002	0.039	0.030
	(2.49)	(3.10)	(2.13)	(4.56)
ρ	-0.049	0.816	-0.466	-0.017
	(-2.63)	(15.6)	(-9.76)	(-3.16)
ϕ_1	-0.169	-0.437	0.721	-
	(-2.65)	(-5.06)	(7.71)	
ϕ_2	-	-0.182	-0.109	-
		(-2.93)	(-1.30)	
ϕ_3	-	0.090	-	-
		(1.80)		
α	-1.126	6.04	4.96	-2.240
	(-9.82)	(8.69)	(0.602)	(-84.5)
σ_v^2	0.024	0.006	0.001	0.001

That provides analyses of the trajectory of the root over the sample period and various episodes are related to the known historical events.

This brief analysis points out the efficacy of filtering application to analyze the path of stochastic unit root in models of real exchange rates. It also establishes the fact that models with fixed rate of mean reversion are likely to induce incorrect inferences about purchasing power parity in high inflation countries.

2.2 Common and Specific Components in Currency Movements

This section is based on the IMF Country Report No. 06/367, published in October 2006. This discussed selected issues on Hungary.

In early and mid-2005, policymakers and market participants were focused on the possible breach of the "strong" edge of Hungary's exchange rate band with the euro. The Czech koruna, the Hungarian forint, the Polish zloty, and the Slovak koruna—the currencies of the four central European (CE-4) new members of the European Union—had appreciated markedly since January 2004. The forint was moving in the range of Ft 241-250 per euro. There was even a concern that speculators may employ strategies to gain from temporary breaches of the strong edge of the band at Ft 240.01 per euro. The Magyar Nemzeti Bank (MNB), in its August 2005 Quarterly Report on Inflation, inferred that markets viewed the prospects of the new members with a high degree of optimism.

The implication was that the country-specific features and fundamentals were overwhelmed by the optimistic view of the region, to the point that markets were willing to discount well known weaknesses, such as Hungary's large fiscal and current account deficits.

Since September 2005, however, the forint has weakened considerably, and its volatility has increased. Between September 2005 and June 2006, the forint weakened from Ft 244 per euro to Ft 282 per euro, a depreciation of 13.5 percent. The forint's volatility also increased, particularly in the months of March and June 2006, when the coefficient of variation (standard deviation/mean) increased to 2 percent and 3 percent, respectively, compared with an average of 0.6 percent in the previous two years. The forint's weakening, especially relative to the other CE-4 currencies, has been attributed, in part, to Hungary's weak fundamentals: the country-specific factors are, apparently, no longer "in the background." The MNB's May 2006 Quarterly Report on Inflation reflected this shift in sentiment.

The spikes in the forint's volatility also coincided with the heightened global emerging market (EM) volatility. The benefit of EU membership proved insufficient to insulate the CE-4 currencies—and particularly the forint—from its own weakness.

This cited IMF report focuses on three objectives. First, it aims to statistically separate the common and country-specific components of CE-4 currency movements. Second, it further decomposes the CE-4's common factor into the part explained by global EM sentiment, and the part that is unique to the CE-4. Finally, it analyzes the trends in these three components, to understand past currency developments, as well as to reflect on near term prospects.

The report decomposes currency fluctuations using two different approaches. The first approach, the state space model, assumes that CE-4 currency movements are formed of two unobserved components: one that is common to all four countries and another that is country specific. These unobserved components are derived using Kalman filter and maximum likelihood estimation techniques. This requires appropriate dynamic specification of these components so that the model could be expressed in state space form.

The second approach, the global factor model, takes the set of global emerging markets and assumes that currency movements are composed of three parts: one that is common to all emerging markets, one that is common to each region, and one that is country specific. The advantage this approach holds over the first model is that it can shed light on the extent to which the co-movement in CE-4 currencies is due to region-specific, as opposed to global, factors.

In this section we only focus on the first of the above two approaches i.e. we discuss the method of decomposition of the currency movements into a common and country specific components. After describing the state space formulation of the problem we will briefly outline the main results reported in the IMF report.

The "unobserved components" dynamic factor analysis used here is based on maximum likelihood Kalman filtering as outlined in chapter 1.

Let it $y_{i,t}$ be the exchange rate of country i at time t, expressed in local currency units per euro, where i = 1 (Czech Republic), 2 (Hungary), 3 (Poland), or 4 (Slovak Republic). There is an unobserved component, κ_t, that is common to all CE-4 currencies, and a second component, $\eta_{i,t}$ that is idiosyncratic or country specific:

$$y_{i,t} = \gamma_i \kappa_t + \eta_{i,t}, \quad i=1,2,3,4 \qquad (2.9)$$

where γ_i is the country specific sensitivity of country I to the common component. This suggests that the impact of the common component may vary from country to country.

Both the common and country-specific unobserved components are assumed to follow autoregressive processes. The persistence of the country-specific factors is allowed to vary across countries:

$$\kappa_t = \phi \kappa_{t-1} + \omega_t \qquad (2.10)$$

$$\eta_{i,t} = \psi_i \eta_{i,t-1} + \varepsilon_{i,t}, \quad i=1,2,3,4 . \qquad (2.11)$$

The innovations ω_t and $\varepsilon_{i,t}$ are assumed to be Gaussian and orthogonal, and the variance of ω_t is normalized to unity in order to achieve identification of the common component:

$$\begin{bmatrix} \omega_t & \varepsilon_{1,t} & \varepsilon_{2,t} & \varepsilon_{3,t} & \varepsilon_{4,t} \end{bmatrix}' \sim N(0,\Sigma) \qquad (2.12)$$

where, $\Sigma = \mathrm{diag}\left(1,\sigma_1^2,\sigma_2^2,\sigma_3^2,\sigma_4^2\right)$.

The system expressed in equations (2.10)–(2.12) can be written in state space form as follows. The measurement equation is,

$$\begin{bmatrix} y_{1,t} \\ y_{2,t} \\ y_{3,t} \\ y_{4,t} \end{bmatrix} = \begin{bmatrix} \gamma_1 & 1 & 0 & 0 & 0 \\ \gamma_2 & 0 & 1 & 0 & 0 \\ \gamma_3 & 0 & 0 & 1 & 0 \\ \gamma_4 & 0 & 0 & 0 & 1 \end{bmatrix} \begin{bmatrix} \kappa_t \\ \eta_{1,t} \\ \eta_{2,t} \\ \eta_{3,t} \\ \eta_{4,t} \end{bmatrix} \qquad (2.13)$$

and the state transition equation is,

$$\begin{bmatrix} \kappa_t \\ \eta_{1,t} \\ \eta_{2,t} \\ \eta_{3,t} \\ \eta_{4,t} \end{bmatrix} = \begin{bmatrix} \phi & 0 & 0 & 0 & 0 \\ 0 & \psi_1 & 0 & 0 & 0 \\ 0 & 0 & \psi_2 & 0 & 0 \\ 0 & 0 & 0 & \psi_3 & 0 \\ 0 & 0 & 0 & 0 & \psi_4 \end{bmatrix} \begin{bmatrix} \kappa_{t-1} \\ \eta_{1,t-1} \\ \eta_{2,t-1} \\ \eta_{3,t-1} \\ \eta_{4,t-1} \end{bmatrix} + \begin{bmatrix} \omega_t \\ \varepsilon_{1,t} \\ \varepsilon_{2,t} \\ \varepsilon_{3,t} \\ \varepsilon_{4,t} \end{bmatrix}. \qquad (2.14)$$

The equations (2.13) and (2.14) represent the state space form of the model being investigated for common and country specific drivers for the currency movements in this block of four countries.

This state space model could be easily estimated using appropriate data. The IMF report suggests that the authors used daily data from the beginning of 2003 to end of June 2006 for the four countries involved. The parameters estimates are given in Table 2.2.

The entries in the above table show that both the common and country specific components are highly persistent, with autoregressive parameters being close to one. This tends to confirm the common perception among researchers that the exchange rates are best modeled as a random walk.

Table 2.2 Parameter Estimates of the State Space Model 2.13 and 2.14.

Autoregressive parameters		
Parameter	Estimates	Standard Error
ϕ	0.999	0.002
ψ_1	0.964	0.016
ψ_2	0.999	0.002
ψ_3	0.985	0.006
ψ_4	0.995	0003
Country Sensitivity		
γ_1	0.061	0.004
γ_2	0.644	0.042
γ_3	0.013	0.001
γ_4	0.071	0.003
Innovation Variance		
$\ln\left(\sigma_1^2\right)$	-4.977	0.052
$\ln\left(\sigma_2^2\right)$	0.122	0.042
$\ln\left(\sigma_3^2\right)$	-7.869	0.051
$\ln\left(\sigma_4^2\right)$	-5.141	0.065

The common component explains a large portion of the variation in CE-4 currencies, but its importance varies across countries, with the Hungarian forint being the outlier. Because the common and country-specific components are not constructed to be orthogonal, they have a nonzero correlation, and perfect variance decomposition is not possible.

However, the importance of the common component may be analyzed in two different ways. First, we perform the variance decomposition as $\text{var}\left(y_{i,t}\right) = \text{var}\left(\gamma_i \kappa_t\right) + \text{var}\left(\eta_{i,t}\right) + \text{cov}\left(\gamma_i \kappa_t, \eta_{i,t}\right)$, but we ignore the covariance term. So, the share of the variance explained by the common component is given by, $\text{var}\left(\gamma_i \kappa_t\right) / \left[\text{var}\left(\gamma_i \kappa_t\right) + \text{var}\left(\eta_{i,t}\right)\right]$.

Second, we regress each CE-4 currency on the common component and a constant, and get the R-squared from the regression.

Table 2.3 gives these two measures of variance decomposition. By both measures, the common component explains a very high proportion of the variation in the Czech koruna, the Polish zloty, and the Slovak koruna. But the common component accounts for a much smaller share of the variance of the Hungarian forint.

The IMF report also suggests that after weakening somewhat in 2003, the common component appreciated steadily from February 2004 until March 2006. This regional appreciation strengthened the Czech koruna by 13 percent, the forint by 11 percent, the zloty by 18 percent and the Slovak koruna by 8 percent during this period.

The country-specific components of the Czech and Slovak koruna have added 2.5 percent and 1.25 percent, respectively, to the total appreciation since 2003. In contrast, country-specific factors have weakened the zloty by 7 percent over the same period. For Hungary, the country-specific component of the forint has weakened by 22 percent over that period.

Table 2.3 Share of Variance Explained by Common Component.

	Variance Share (First Method)	Variance Share (R-Squared)
Czech koruna	0.97	0.97
Hungarian forint	0.45	0.01
Polish zloty	0.90	0.90
Slovak koruna	0.86	0.84

2.3 Persistent in Real Interest Rate Differentials

This section is based on the article by Cavaglia (1992). This research attempts to explore the questions: Are real interest rates indeed different across countries, and if so, is there a tendency for them to return to equality?

Policy makers, in particular, would also be interested in the length of time it takes for real interest rates to be equalized. Insofar as government policies can affect domestic real interest rates, the persistence of real interest differentials would provide a measure of the extent to which national authorities can exercise independent influence over their domestic financial markets.

Most theoretical models in the open macroeconomics literature make some explicit or implicit assumption about real interest differentials across countries. The perfect co-movement of national real interest rates is one of the primary mechanisms through which Frenkel and Razin (1986) obtain the negative covariance of consumption patterns across countries. The analysis of movements in real interest differentials is also useful for understanding the time series properties of real exchange rates. This can be explained by starting from the interest parity arbitrage condition. This arbitrage condition states that the return on an investment in a foreign asset, which is covered by a forward exchange contract, will yield the same return as an investment in a domestic asset of the same risk class.

We give below details of model formulation following Cavaglia (1992). The interest rate parity arbitrage implies,

$$f(t, t+n) - s_t = i^*(t, t+n) - i(t, t+n) \qquad (2.15)$$

where, s_t is log of the spot exchange rate (foreign currency price of one unit of domestic currency), $f(t, t+n)$ is log of the forward exchange rate at t for delivery at $t+n$. Again, $i(t, t+n)$ is the nominal domestic interest rate for the period t to $t+n$. Similarly, i^* is the foreign interest rate.

Forward exchange rate has been decomposed into two part – expected future exchange rate and a risk premium term as in Wolff (1987). This suggests,

$$s^e(t, t+n) + k^e(t, t+n) - s_t = i^*(t, t+n) - i(t, t+n) \qquad (2.16)$$

where the terms with superscript 'e' implies expected values and 'k' represent foreign exchange risk premium.

It is now convenient to express (2.16) in real term using Fisher hypothesis. Thus,

$$q(t) = r^e(t,t+n) - r^{e*}(t,t+n) + k^e(t,t+n) + q^e(t,t+n) \quad (2.17)$$

where $p(t)$ is the logarithm of the price level at t, $p^e(t,t+n)$ is the expected logarithm of the price level at t+n, $q(t) - s(t) + p(t) - p^*(t)$, $q^e(t,t+n) = s^e(t,t+n) + p^e(t,t+n) - p^{e*}(t,t+n)$ and the ex-ante real interest rate between t and t+n seen at t is $r^e(t,t+n)$.

If we assume that the expectation theory of interest rate applies then (2.17) may be written as,

$$q(t) - q^e(t,t+n) - k^e(t,t+n)$$
$$= \frac{1}{n} \sum_{j=0}^{n-1} r^{e*}(t+j,t+j+1) - r^e(t+j,t+j+1). \quad (2.18)$$

There is an economic interpretation to the equation (2.18). To the extent that the premium term is small and constant, analysis of the time series properties of real interest differentials helps us to understand the time series properties of real exchange rates and the speed at which they are expected to change towards their value n periods into the future.

When n is large, this is equivalent to studying the speed at which exchange rates are expected to change towards their long-run expected value. A high variability of the interest differential also suggests a large variability of real exchange rates around their long-run values.

Thus, the characterization of the time series properties of ex ante real interest differentials provides information regarding the time series properties of real exchange rates as well. This gives us a statistical basis for interpreting of models of economic interdependence. Existing studies on real interest rates do not necessarily address whether there is a long-run tendency for real interest rates to be equalized over time. Cavaglia (1992) attempts to address this point and adopts a state space approach to modeling ex ante real interest rates. In this framework,

$$D^*(t,t+1) = r^e(t,t+1) - r^{e*}(t,t+1) \quad (2.19)$$

where, $D^c(t,t+1)$ represents the ex ante real interest rate differential from time t to t+1, seen at time t. $D(t,t+1)$ is the ex post real interest rate differential seen at time t+1. Thus,

$$D(t,t+1) = r(t,t+1) - r^*(t,t+1),$$
$$E(t,t+1) = \varepsilon(t,t+1) - \varepsilon^*(t,t+1).$$

(2.20)

Here, $E(t,t+1)$ is the cross country differential in inflation forecast errors, observed at t+1. If we assume that the ex ante real interest rate differential follow an autoregressive process of order p, then the following state space model results:

$$D(t,t+1) = D^c(t,t+1) + E(t,t+1)$$

(2.21)

$$D^c(t+1,t+2) = \phi(L).D^c(t,t+1) + u(t+1)$$

(2.22)

with $\phi(L) = \phi_1 + \phi_2 L + ... + \phi_p L^{p-1}$. In the state space terminology, (2.21) is the measurement equation and (2.22) is state transition equation. Obviously, estimation of such a model can be achieved by Linear Kalman Filter described in chapter 1. Cavaglia (1992) applies standard time series technique to arrive at the order p of the autoregressive process and assumes that the covariance matrix is time homogeneous.

Data used in Cavaglia study consist of one-month Euro deposit rate for the US, Germany, Switzerland and Holland covering the period 1973 to 1979. Nominal interest rates were deflated by domestic inflation rates computed from the consumer price index reported by the IMF. The parameter estimates were carried out by likelihood maximization as discussed in chapter 1 Additional computational details could be obtained from Cavaglia (1992).

Table 2.4 gives the estimation results for the four countries and both the US and Germany were considers as the reference countries to compute the interest rate differentials. The selected models for the ex ante real interest differentials are all mean zero stationary processes. This is evidence in support of theoretical models of economic interdependence which imply a tendency towards real rate equality. Deviations from real rate equality are short-lived. The models presented in Table 2.4 can be utilized to generate forecasts of expected short-term real interest differentials.

Table 2.4 Selected Time Series Models for Ex Ante Real Interest Rates.

US - Germany

$$\left(1 - 0.502L - 0.225L^{12}\right) D^e \left(t, t+1\right) = u\left(t\right)$$

US – Switzerland

$$\left(1 - 0.415L - 0.133L^6 - 0.248L^{12}\right) D^e \left(t, t+1\right) = u\left(t\right)$$

US – Holland

$$\left(1 - 0.175L - 0.299L^6 - 0.317L^{12}\right) D^e \left(t, t+1\right) = u\left(t\right)$$

Germany – Switzerland

$$\left(1 - 0.129L - 0.135L^2 - 0.617L^{12}\right) D^e \left(t, t+1\right) = u\left(t\right)$$

Germany – Holland

$$\left(1 - 0.215L + 0.149L^2 - 0.709L^{12}\right) D^e \left(t, t+1\right) = u\left(t\right)$$

It was also argued that, to the extent that exchange rate risk premia and term structure premia are small, the time series properties of ex ante differentials provide information about deviations of exchange rates from their long-run value. On the whole, it was found that these deviations for European currencies relative to the U.S. dollar were more variable than those relative to the German Deutschmark.

In Table 2.4 only the statistically significant parameters have been shown.

2.4 Risk Premia in Forward Exchange Rate

The efficacy of forward exchange rate as an unbiased predictor of future spot exchange rate has been the subject of research for many years. However, observed deviation from this hypothesis has been attributed to the existence of risk premium in the forward exchange rate. In this section we review two methodologies proposed in the literature for inferring this unobserved risk premium. Both these approaches rely on signal extraction mechanism with a basic difference in setting up the framework. The more recent approach uses the derivatives pricing principle that relates the historical and the risk neutral measures with a suitable specification of the market price of risk. The earlier approach directly specifies the risk premium in the historical measure. We compare these two methods in term of their ability to predict spot exchange rate one step ahead and contrast these with that of a random walk forecast.

Several regression based studies attempted to explain the ability (or otherwise) of the forward exchange rates in predicting the future realized spot exchange rates. Although with the improvements in econometric theory the nature of the tests employed have changed, but the basic approach has remained essentially within the regression framework. For example, Wu and Zhang (1997) employ a non-parametric test and not only reject the unbiasedness hypothesis but also conclude that the forward premium either contains no information or wrong information about the future currency depreciation. On the other hand, Bakshi and Naka (1997) derive an error correction model under the assumption that the spot and the forward rates are co-integrated and conclude using the generalized method of moments that the unbiasedness hypothesis cannot be rejected. Phillips and McFarland (1997) develop a robust test and reject the unbiasedness hypothesis but conclude that the forward rate has an important role as a predictor of the future spot rate.

The failure of the unbiasedness hypothesis has been attributed to the existence of a foreign exchange risk premium. This has led to a great deal of research on the modeling of the risk premia in the forward exchange rate market. However, models of risk premia have been unsuccessful in explaining the magnitude of the failure of unbiasedness

(Engel (1996), page 124). We define the term $rp_t \equiv f_{t,k} - E_t\left(s_{t+k}\right)$ as the foreign exchange risk premium. Under risk-neutrality the market participants would behave in such a way that $f_{t,k}$ equals $E_t\left(s_{t+k}\right)$ and the expected profit from forward market speculation would be zero. Stulz (1994) discusses a model of foreign exchange risk premium based on optimizing behavior of international investors. However, alongside such theoretical developments pure time series studies of rp_t have also assumed a renewed importance. These are useful in describing the behavior of $f_{t,k} - E_t\left(s_{t+k}\right)$. Models of the foreign exchange risks premium that assume rational expectations should be able to explain the observed time series properties. Examples of such studies include Backus et al (1993) and Bekaert (1994).

Modeling of the time varying risk premia has been inadequately addressed in the literature since there is little theory to guide us in this respect. Wolff (1987) and Cheung (1993) have modeled the risk premia as an unobserved component and estimated it using the Kalman filter. In their signal extraction approach they empirically determine the temporal behavior of the risk premium using only data of forward exchange rate and the spot exchange rate. Although the signal extraction approach avoids specifying any particular form of the risk premia, it offers little insight into the risk premia and other economic variables. Cheung (1993) links the estimated risk premia with other macro economic variables. This relies on the inter-temporal asset-pricing model of Lucas (1982). However, the results are not very encouraging and the estimated regression models have very low R-squares.

Both Wolff (1987) and Cheung (1993) analyze the quantity $\left(f_{t,k} - s_{t+k}\right)$ to determine the time series characteristic of the unobserved risk premia. This in turn determines the dynamics of the unobserved component, the risk premia. For different currencies they examine the dynamics of the risk premia can be captured by a low order ARMA process. Wolff (2000) further extends the number of currencies studied in the same framework. In those papers, therefore, the observed difference between the forward exchange rate at time t for the period t+k and the subsequently realized spot rate at time t+k is the main driver for the structure of the risk premia. This is assumed to be composed of the unobserved risk premia and the unexpected depreciation of the exchange rate. The complete model is, thus, specified in the historical measure.

Bhar, Chiarella and Pham (2001) propose an alternative approach to this problem of unobserved risk premium. They model the market price of risk (and hence the risk premia) by utilizing the no-arbitrage relation between the spot and the forward markets. This requires an assumption about the dynamic process for the market price of risk. Then the dynamic system involving forward rate, spot rate, interest rates and the market price of risk could be cast into the state space model framework. The system is then estimated by Kalman filter. This approach fundamentally differs from those of Wolff (1987) and Cheung (1993) in that it models the spot and forward dynamics as well as the market price of risk which relates the historical and the risk-neutral measures. This approach can also be extended to foreign exchange options, which are a rich source of untapped information about markets' view of risk premia.

Both these approaches to estimating forward exchange rate risk premium are applied to a set of exchange rate data and the results compared. Since both these models rely of filtering and one step ahead prediction, it is possible to compare this prediction of spot exchange rate with a random walk model. Both these approaches predict one step ahead spot exchange rate better than a random walk model.

2.4.1 Approach based on Market Price of Risk (BCP)

Let the spot exchange rate follow the one-dimensional geometric diffusion process,

$$dS = \mu S dt + \sigma_S S dW(t) \tag{2.23}$$

where μ is the expected return from the spot asset, σ_S is the volatility of this return, both measured per unit of time and dW is the increment of a Wiener process under the so-called historical (statistical) probability measure Q, r is the domestic risk-free interest rate and r_f as the counterpart in the foreign currency. Since r_f can be interpreted as a continuous dividend yield, the instantaneous return to an investor holding foreign exchange is $(\mu + r_f)$. Thus the relationship between the excess return demanded and the market price of risk (λ) may be written

$$(\mu + r_f) - r = \lambda\sigma_S, \; \mu = (r - r_f) + \lambda\sigma_S . \tag{2.24}$$

Thus, under the historical measure Q equation (2.23) can be rewritten

$$dS = (r - r_f + \lambda\sigma)Sdt + \sigma_S SdW(t) . \tag{2.25}$$

Alternatively under the risk neutral measure $\tilde{Q}$ the last equation becomes

$$dS = (r - r_f)Sdt + \sigma_S Sd\tilde{W}(t), \; \tilde{W}(t) - W(t) \mid \int_0^t \lambda(u)du . \tag{2.26}$$

We recall that under $\tilde{Q}$, the process $\tilde{W}(t)$ is not a standard Wiener process since $E[d\tilde{W}(t)] = \lambda dt \neq 0$ in general. However, Girsanov's theorem allows us to obtain the equivalent measure $\tilde{Q}$ under which $\tilde{W}(t)$ does become a standard Wiener process. The measures Q and $\tilde{Q}$ are related via the Radon-Nikodym derivative.

Using standard arguments for pricing derivative securities (see for example, Hull (1997), chapter 13), the forward price at time t for a contract maturing at T (>t), is

$$F(t,T) = \tilde{E}_t(S_T) . \tag{2.27}$$

But from equation (2.26),

$$d[S(t)e^{-(r-r_f)t}] = \sigma_S S(t)e^{-(r-r_f)t}d\tilde{W}(t) \tag{2.28}$$

so that under $\tilde{Q}$, the quantity $S(t)e^{-(r-r_f)t}$ is a martingale and it follows immediately that,

$$\tilde{E}_t(S_T) = S_t e^{(r-r_f)(T-t)}, \; F(t,T) = S_t e^{(r-r_f)(T-t)} . \tag{2.29}$$

If the maturity date of the contract is a constant period, x, ahead then (2.28) may be written as

$$F(t,t+x) = S_t e^{(r-r_f)x} . \tag{2.30}$$

Then from (2.25), (2.26) and (2.30) and by a trivial application of Ito's lemma we obtain the stochastic differential equation for F under Q and $\tilde{Q}$. Thus, under $\tilde{Q}$

$$dF(t,x) = (r - r_f)F(t,x)dt + \sigma_S F(t,x)d\tilde{W}(t) \tag{2.31}$$

and under Q,

$$dF(t,x) = (r - r_f + \lambda\sigma_S)F(t,x)dt + \sigma_S F(t,x)dW(t) . \tag{2.32}$$

with $F(0,x) = S_0 e^{(r-r_f)x}$.

We now assume that under historical measure Q the market price of risk, λ , follows the mean reverting stochastic process,

$$d\lambda = \kappa(\overline{\lambda} - \lambda)dt + \sigma_\lambda dW \qquad (2.33)$$

where $\overline{\lambda}$ is the long-term average of the market price risk, κ defines the speed of mean reversion. Here, we assume that the same noise process drives both the spot exchange rate and the market price of risk. It would of course also be possible to consider a second independent Wiener process driving the stochastic differential equation for λ . However, we leave investigation of this issue for future research.

It should be pointed out here that when discretised the stochastic differential equation (2.33) would become a low order ARMA type process of the kind reported in Wolff (1987) and Cheung (1993). The parameters in equation (2.33) may be estimated from the data using the Kalman filter as pointed out earlier.

Considering we have one forward price, $f(t,x)$, then we have a system of 3 stochastic differential equations. These are (under the measure Q),

$$dS = (r - r_f + \lambda\sigma_S)Sdt + \sigma_S SdW(t) \qquad (2.34)$$

$$d\lambda = \kappa(\overline{\lambda} - \lambda)dt + \sigma_\lambda dW(t) \qquad (2.35)$$

$$dF(t,x) = (r - r_f + \lambda\sigma_S)F(t,x)dt + \sigma_S F(t,x)dW(t) \qquad (2.36)$$

where, $S(0) = S_0$, $\lambda(0) = \lambda_0$, $f(0,x) = S_0 e^{(r-r_f)x}$.

It should be noted that the information contained in equations (2.36) is also contained in the pricing relationships,

$$F(t,x) = S_t e^{(r-r_f)x} . \qquad (2.37)$$

To estimate the parameters in the filtering framework, however, it is convenient to work with the equation (2.36).

From equation (2.25), we can write the spot price at time $t + x$ as, using $s(t) = \ln S(t)$, as

$$s(t+x) = s(t) + \left(r - r_f - \frac{\sigma_S^2}{2}\right)x + \sigma_S \int_t^{t+x} \lambda(\tau)d\tau + \sigma_S \int_t^{t+x} dW(\tau) . \qquad (2.38)$$

From equation (2.38) we can write the expected value of $s(t+x)$ as

$$E_t\left[s(t+x)\right] = s_t + \left(r - r_f - \frac{\sigma_S^2}{2}\right)x + \sigma_S E_t\left[\int_t^{t+x} \lambda(\tau)d\tau\right]. \quad (2.39)$$

The derivation of this expectation may be found in Bhar, Chiarella and Pham (2001). It turns out to be:

$$E_t\left[s(t+x)\right] = s(t) + \left(r - r_f - \frac{\sigma_S^2}{2}\right)x$$

$$+ \sigma_S\left[\left(\lambda(t) - \overline{\lambda}\right)\left(\frac{1-e^{-\kappa x}}{\kappa}\right) + \overline{\lambda}x\right]. \quad (2.40)$$

The above equation may also be expressed (via use of equation (2.30) as,

$$E_t\left[s(t+x)\right] = f(t,x) - \frac{\sigma_S^2}{2} + \sigma_S\left[\left(\lambda(t) - \overline{\lambda}\right)\left(\frac{1-e^{-\kappa x}}{\kappa}\right) + \overline{\lambda}x\right]. \quad (2.41)$$

Let $\pi(t,x)$ represent the risk premium (under Q) for the x period ahead spot rate, then from equation (2.41)

$$\pi(t,x) = -\frac{\sigma_S^2}{2} + \sigma_S\left[\left(\lambda(t) - \overline{\lambda}\right)\left(\frac{1-e^{-\kappa x}}{\kappa}\right) + \overline{\lambda}x\right]. \quad (2.42)$$

It has been pointed out that previous studies attributed the difference between the forward rate and the subsequently realized spot exchange rate to a risk premium and the unexpected depreciation of the exchange rate. Equation (2.42) gives an explicit expression for the risk premium, characterizing how the market price of risk enters the expectation formation and thus influence the risk premium. The integral terms involving the Wiener increments in equation (2.38) may be related to the noise terms identified in Wolff (1987) and Cheung (1993).

In order to compute the time variation of risk premia for one-month forward rates obtained from equation (2.42) for a given exchange rate, we need the estimates of the parameters describing the stochastic process for λ given by equation (2.33). Bhar, Chiarella and Pham (BCP) (2001) show the state space formulation of the system and estimation of these parameters as well as the filtered and smoothed estimates of $\lambda(t)$. They also point out that this method can be easily applied to multiple forward exchange rates and thereby examine the term structure of forward risk premia present in the quoted forward exchange rates.

In this section, we would also like to compare this with the risk premia obtained from the approach outlined in Wolff (1987) and Cheung (1993). Both these methods require data synchronization with respect to matching the forward rate period. This is not necessary in the method just outlined above.

In the following sub-section we also briefly describe the method of Wolff/Cheung so as to facilitate comparison with the above approach.

2.4.2 Method of Wolff/Cheung

The main idea in modeling the forward exchange rate risk premia in their model is the assumption that the forecast error resulting from the forward rate as the predictor of futures spot exchange rate consists of a premium component and a white noise error. In this context,

$$S_{t+1} = F_{t,t+1} + P_t + \varepsilon_t . \qquad (2.43)$$

Here ε_t is an uncorrelated zero-mean sequence and P_t is the unobserved risk premium. In terms of state space system representation this is the measurement equation. Both Wolff and Cheung determine the dynamic of the risk premium by studying the time series properties (in the Box-Jenkins sense) of the quantity $\left(F_{t,t+1} - S_{t+1}\right)$. As suggested in Wolff (2000) for most currencies either an ARMA (1,1) or an MA(1) representation is adequate. The corresponding equations for P_t are given by,

$$P_t = \phi P_{t-1} + \vartheta_t , \text{ for AR(1)} , \qquad (2.44)$$

$$P_t = \phi P_{t-1} + \theta \vartheta_{t-1} + \vartheta_t , \text{ for ARMA(1,1)} , \qquad (2.45)$$

$$P_t = \theta \vartheta_{t-1} + \vartheta_t , \text{ for MA(1)} . \qquad (2.46)$$

It is straightforward to set up the state transition equation matrices with the above assumed dynamics of the unobserved risk premium. Assuming v^2 is the variance of the innovation ϑ the state space representation is given below. The estimation algorithm for these models is straightforward linear Kalman filter and same as for Bhar, Chiarella and Pham (2001) framework.

For AR (1) specification the measurement equation is (2.43) and the state equation is (2.44). For ARMA (1,1) specification the state transition equation is:

$$\begin{bmatrix} P_t \\ \theta\vartheta_t \end{bmatrix} = \begin{bmatrix} \phi & 1 \\ 0 & 0 \end{bmatrix} \begin{bmatrix} P_{t-1} \\ \theta\vartheta_{t-1} \end{bmatrix} + \begin{bmatrix} 1 \\ \theta \end{bmatrix} \vartheta_t \qquad (2.47)$$

and the measurement equation takes the form:

$$S_{t+1} = F_{t,t+1} + \begin{bmatrix} 1 & 0 \end{bmatrix} \begin{bmatrix} P_t \\ \theta\vartheta_t \end{bmatrix} + \varepsilon_t . \qquad (2.48)$$

For the MA(1) representation the matrices are similar to that in case of ARMA(1,1) with restriction that $\phi = 0$.

2.4.3 Data and Empirical Results

As part of our empirical investigation we apply both the methodologies reviewed here to five different exchange rates, all against U.S. dollars. We restrict ourselves to the pre-Euro era. These are Australian dollar (AUD), German marks (DEM), French frank (FRF), British pound (GBP) and the Japanese yen (JPY). The data set covers the period January 1986 to December 1998 with 156 observations in each series. We use only the one-month forward exchange rates so that the results from the BCP model can be compared directly with those from the implementation of Wolff/Cheung methodology. It should be pointed out that the BCP methodology does not require that the observations should be properly aligned with the maturity of the forward rates. The exchange rate data reflects the daily 4PM London quotation obtained from DataStream™ and the interest rate data are the daily closing one-month Euro currency deposit rates.

To start the adaptive algorithm of the Kalman filter we initialize the state vector with the first observations. The algorithm also requires specifying the prior covariance matrix for the state vector. In the absence of any specific knowledge about the prior distribution we use the diffuse prior specification following Harvey (1989, p. 121).

The parameter estimates are obtained by maximizing the log likelihood function discussed in detail in BCP (2001). The numerical optimization algorithm called 'Newton' in GAUSS™ is used for this purpose without any parameter constraints. The results of the estimation procedure are shown in Table 2.5. The t-statistics reported in that table are computed from the standard error obtained from the heteroscedasticity consistent covariance matrix of the parameters at the point of convergence.

All the parameters of the model except the long-term average market price of risk are statistically significant for each of the currencies. The estimated parameter σ_S compares favorably with the sample estimates obtained from the spot exchange rate series (not reported separately).

How the model fits the data is best analyzed by examining the residual from the estimation process. These are reported in Table 2.6. One of the main requirements is that the residual be serially uncorrelated both in its level and its squared form. The portmanteau test and the ARCH test support this requirement for all the currencies examined. As the Kalman filter generated residuals are recursive in nature two other tests are carried out to judge the model adequacy. These are modified Von Neumann ratio and the recursive t-tests (see Harvey (1990), page 157). Both these tests support our modeling approach. Finally, the conditional normality assumption made in the modeling is also supported by the Kolmogorov-Smirnov test statistics reported in Table 2.6.

Since we are interested in comparing and contrasting the two modeling approaches we present the estimation results of the Wolff/Cheung models for the same set of currencies in Table 2.7. As can be seen most parameters are statistically significant. We subject the model residuals to the same set of tests as in Table 2.6 (although in their original papers, Wolff/Cheung does not report these diagnostics). The results reported in Table 2.8 support all the model adequacy tests.

For the BCP model the risk premia contained in the one-month forward exchange rate can be computed easily from equation (2.42) with the help of the estimated parameters from Table 2.5 and the filtered (or smoothed) estimates of the market price of risk. Since Wolff/Cheung method does not provide this risk premia directly we dot not analyze this aspect any further.

Next, we compare the one-month ahead prediction of the spot exchange rate with the realized exchange rate and thus generate the mean absolute prediction error and the root mean squared prediction error for each of the exchange rate series. In the context of Kalman filter this is really ex ante prediction error since the prediction of the measurement variable for time t+1 is made utilizing information up to and including time t. This is true for Wolff/Cheung model as well because the way we have implemented it. The comparative results are shown in Table 2.9 for our model, Wolff/Cheung model as well as a martingale process.

Overall conclusion from examining the Table 2.9 is that both the models perform better than the martingale process. There is, however, not much difference in forecasting performance between the BCP model and Wolff/Cheung model. It should, however, be remembered that the BCP model could be implemented for data set of any observed frequency whereas Wolff/Cheung approach is limited to data set where the spot exchange rate frequency aligns with the forward rate data used.

2.4.4 Summary of Section 2.4

In this section we have reviewed two approaches to analyze the time varying risk premium in forward exchange rates. The first one, the BCP model exploits the relationship that links the spot exchange rate and the forward exchange rate through the market price of risk. By directly modeling the market price of risk as a mean reverting process it is possible to show how the market price of risk enters into expectation formation for a future spot exchange rate.

This BCP methodology allows us to quantify the risk premium associated with a particular forward exchange rate in terms of the parameters of the process describing the market price of risk. We also discuss how these parameters can be estimated in a state space framework by application of Kalman filter. This procedure, in turn, generates the filtered and the smoothed estimates for the unobserved market price of risk.

The second methodology i.e. Wolff/Cheung procedure is also applied to the same data set and for three different dynamics of the unobserved risk premium process. This model is also estimated via Kalman filter.

Both these models perform well in one step ahead forecasting spot exchange rate and both beat the naïve forecast. However, the BCP model is well suited for any observed data frequency, whereas in the Wolff/Cheung method spot and forward exchange rate data frequency needs to be matched.

Table 2.5 Parameter Estimates of Market Price of Risk Model.

	σ_S	κ	λ	σ_λ
AUD	0.0272	8.0884	0.7226	21.2137
	(2.16)	(2.81)	(0.68)	(2.16)
DEM	0.0797	17.8281	-0.2457	6.0437
	(9.48)	(7.88)	(-0.67)	(2.98)
FRF	0.0528	11.2577	-0.5531	13.2373
	(3.45)	(5.97)	(-0.80)	(2.58)
GBP	0.0609	12.2235	-0.6545	11.8079
	(5.12)	(5.85)	(-1.12)	(4.54)
JPY	0.0960	9.3827	0.0590	4.3061
	(6.79)	(0.98)	(0.27)	(2.15)

Numbers in parentheses are t-statistics computed from standard errors obtained using the heteroscedasticity consistent covariance matrix at the point of convergence.

Table 2.6 Residual Diagnostics and Model Adequacy Tests (Model in Table 2.5).

Market	Portmanteau	ARCH	MNR	Recursive T	KS Statistic
AUD	0.226	0.702	0.832	0.597	0.042
DEM	0.080	0.474	0.996	0.887	0.055
FRFR	0.482	0.494	0.871	0.917	0.082
GBP	0.091	0.342	0.897	0.857	0.068
JPY	0.286	0.608	0.600	0.956	0.064

Entries are p-values for the respective statistics except for the KS statistic. These diagnostics are computed from the recursive residual of the measurement equation, which corresponds to the spot index process. The null hypothesis in portmanteau test is that the residuals are serially uncorrelated. The ARCH test checks for no serial correlations in the squared residual up to lag 26. Both these test are applicable to recursive residuals as explained in Wells (1996, page 27). MNR is the modified Von Neumann ratio test using recursive residual for model adequacy (see Harvey (1990, chapter 5). Similarly, if the model is correctly specified then Recursive T has a Student's t-distribution (see Harvey (1990, page 157). KS statistic represents the Kolmogorov-Smirnov test statistic for normality. 95% significance level in this test is 0.109. When KS statistic is less than 0.109 the null hypothesis of normality cannot be rejected.

Table 2.7 Parameter Estimates for the Model Based on Wolff (1986), Cheung (1993).

	ϕ	θ	ε	ν
AUD	0.9439		0.0261	0.0000
	(23.18)		(17.01)	(0.000)
DEM	-0.7150	0.9206	0.0142	0.0277
	(-6.57)	(5.12)	(1.64)	(4.28)
FRF	-0.7183	0.9189	0.0077	0.0288
	(-7.40)	(9.70)	(5.56)	(15.31)
GBP		0.6318	0.0269	0.0159
		(5.90)	(6.14)	(2.25)
JPY		0.9311	0.0368	0.0054
		(3.91)	(10.27)	(0.57)

Numbers in parentheses are t-statistics computed from standard errors obtained using the heteroscedasticity consistent covariance matrix at the point of convergence.

Table 2.8 Residual Diagnostics and Model Adequacy Tests (Model in Table 2.7).

Market	Portmanteau	ARCH	MNR	Recursive T	KS Statistic
AUD	0.130	0.769	0.297	0.925	0.035
DEM	0.428	0.938	0.604	0.482	0.055
FRFR	0.591	0.937	0.379	0.275	0.059
GBP	0.270	0.420	0.486	0.287	0.083
JPY	0.458	0.551	0.539	0.942	0.063

Entries are p-values for the respective statistics except for the KS statistic. These diagnostics are computed from the recursive residual of the measurement equation, which corresponds to the spot index process. The null hypothesis in portmanteau test is that the residuals are serially uncorrelated. The ARCH test checks for no serial correlations in the squared residual up to lag 26. Both these test are applicable to recursive residuals as explained in Wells (1996, page 27). MNR is the modified Von Neumann ratio test using recursive residual for model adequacy (see Harvey (1990, chapter 5). Similarly, if the model is correctly specified then Recursive T has a Student's t-distribution (see Harvey (1990, page 157). KS statistic represents the Kolmogorov-Smirnov test statistic for normality. 95% significance level in this test is 0.109. When KS statistic is less than 0.109 the null hypothesis of normality cannot be rejected.

Table 2.9 One Step Ahead Forecast Error for Spot Exchange Rate.

	Model Based On	MAE	MSE
AUD	Market Price of Risk	0.0205	0.0007
	Cheung/Wolff	0.0206	0.0007
	Martingale Process	0.0279	0.0013
DEM	Market Price of Risk	0.0258	0.0011
	Cheung/Wolff	0.0254	0.0010
	Martingale Process	0.0451	0.0035
FRF	Market Price of Risk	0.0248	0.0010
	Cheung/Wolff	0.0241	0.0009
	Martingale Process	0.1446	0.0344
GBP	Market Price of Risk	0.0235	0.0010
	Cheung/Wolff	0.0250	0.0011
	Martingale Process	0.0143	0.0004
JPY	Market Price of Risk	0.0286	0.0014
	Cheung/Wolff	0.0290	0.0014
	Martingale Process	3.6540	23.5860

'MAE' and 'MSE' represent mean absolute error and mean squared error respectively. These are computed from the one step ahead forecast error obtained during Kalman filter recursion. These forecast errors are used to develop the prediction error form of the likelihood function. Cheung/Wolff model refers to our somewhat modified implementation of their approach.

2.5 Concluding Remarks

This chapter focused on filtering applications in the foreign exchange market on four different dimensions. First, it showed how filtering concepts ate applied to extract information on mean reversion rates in real exchange rates. The information on the time trajectory of the mean reversion rate is highly informative economically. Second, it focused on identifying common and specific components of exchange rate movements for a similar group of countries from the same geographical region. Third, the discussion centered on understanding the behavior of ex ante real interest rate and how it is related to the exchange rate risk premia. Finally, the section discussed on signal extraction methods of inferring unobserved risk premia inherent in the forward exchange rates.

Chapter 3

Equity Market – Filtering Applications

Equity market risk premium is a very fiercely researched topic. There is good reason for that in the sense that its perception alters investors' behavior. At the same time it is not a directly observable quantity. It is thought that the risk premium is a product of two components – amount of risk and the unit price of risk. At the same time the literature in financial economics is full with various ways to estimate the level of risk in the market. However, the unit price of risk remains an unobserved quantity and requires stochastic filtering to be extracted. In this chapter we deal with two aspects of this issue. In the first instance, we show how to infer the unobserved unit price or risk and the related economic notion of co-movement among several equity markets. Once the unit price of risk is estimated, it is employed in measuring the time path of economic convergence in comparable economies. So far, the notion of economic convergence has been measured by using economic variables like gross domestic product. However, our contention is that economic convergence should imply convergence in price of risk in the two economies. We again employ filtering technique to quantify the time path of this convergence via the price of risk.

In the second instance, we develop a model for estimating ex-ante risk premium using information on dividend yield and earnings yield from the market. All these relevant variables are assumed to follow different stochastic processes and the unobserved quantity that bind these processes, the ex ante risk premium, is inferred via a filtering system.

3.1 Introduction to Equity Price of Risk

A thorough understanding of the nature of the risk premium in asset markets is an integral part of financial decision making. This is particularly so in equity markets. Most researchers have focussed on the beta risk of stocks or stock portfolios in different experimental settings. The beta is now well established to be time-varying, and different models have been proposed to capture its time variation. Groenewold and Fraser (1999) find significant structural breaks in betas using data from the Australian industry sector. They capture the time variation of the betas using three different approaches: recursive regression, rolling regression, and the Kalman filter. Faff, Hillier, and Hillier (2000) suggest that if the Kalman-filter-based approach incorporated the information contained in the GARCH (Generalized Autoregressive Heteroscedasticity) type models, it would provide a more efficient estimate of the time-varying systematic risk. Choudhry (2002) employs an MA (Moving Average)-GARCH type model to capture the time-varying beta for several UK companies. In doing so, he investigates the stochastic structure of the betas and finds that most are covariance stationary and mean-reverting.

Merton (1980) provides the basis for an alternative approach to the analyses of the stock market risk premium. Merton's model of inter-temporal capital asset pricing demonstrates that a risk-averse investor demands a risk premium to compensate for the market risk. Assuming that the variance of the market return is a sufficient statistic for the market risk, this risk premium is proportional to the market risk. In the case of a representative investor, this proportionality constant is also equal to the investor's relative risk aversion. This multiplier can also be used to measure risk by providing the unit price of risk. There is no evidence to suggest that the unit price of risk should remain constant over time. This, in fact, is the approach adopted by Jochum (1999). Jochum shows that the time variation in the price of risk can be estimated using an unobserved component model that also allows the measure of risk to vary with time as an ARCH process. Chou, Engle, and Kane (1992) apply a similar methodology and relate the inferred price of risk to the risk aversion parameter of the investor.

The notion of time variation of the unit price of risk is by no means limited to the framework outlined above. The derivative contracts written on the market indices of many countries offer another potential source of data for inferring the time variation of the market price of risk. The concept is embedded in the rich theory of derivatives pricing, a theory which requires a relation between risk-neutrality and historical probability distributions. Bhar, Chiarella, and Runggaldier (2004) demonstrate how the model can be set up and estimated using the state-space structures. This approach allows us to infer the time variation of the market price of risk quite efficiently.

This section adopts the concept of time variation in the price of risk and applies it to the G7 markets. Our methodology is similar to that of Jochum (1999) and can be used without relying on complete data from the derivatives markets of all countries over long periods. The price-of-risk series inferred from our model thus allows us to further investigate the comparative behaviour between these markets.

World stock markets have become more integrated over the last three decades, chiefly as a consequence of financial deregulation and advances in computer technology. Financial researchers have traced the evolution of this trend from different perspectives, applying diverse methodologies and databases to elucidate the linkage phenomenon. Fundamentally, however, most have taken the common approach of modelling return series from similar asset markets and analysing feedback mechanisms through the first or the second moments of the distributions. Malliaris and Urrutia (1992, 1997) show that certain global events tend to move national stock markets in unison. McCarthy and Najand (1995) also provide a good account of the literature in stock market linkages.

After inferring the price of risk, we extend the analysis of stock market linkages to a deeper level by investigating the co-movement of the market price of risk in these G7 markets using concordance, a non-parametric test recently suggested by Harding and Pagan (1999). The concordance measure has been extended by McDermott and Scott (1999) for its distributional properties and successfully applied in studies of co-movement of prices in seemingly unrelated commodities (e.g., Cashin, McDermott, and Scott, 1999). This approach lets us explore whether investors' perceptions of the price of risk move together in these markets over the sample period. In doing so, the notion of total risk premium in the market is divided into two components: the measure of risk and the price of risk. Through this approach we can study the stock market linkages using more sensitive information than the equity return and/or return volatility itself.

If the equity markets are financially integrated, the price of risk should be the same across markets. The price of risk may differ across markets if barriers to capital flows across markets or other influences impede the financial integration of the markets. This is what compels us to investigate the price-of-risk behaviour in several economies. We find that the measure tends to change over time in a pattern consistent with growing levels of integration.

3.1.1 A Model for Equity Price of Risk

The model used in this paper is similar to that of Chou, Engle, and Kane (1992). It can be expressed by the following equation (3.1). If x_t represents the excess return on the equity index at time t, σ_t^2 represents the conditional variance of the excess return at time t, and b_t denotes the price of unit risk, then

$$x_t = b_t \sigma_t^2 + \varepsilon_t \qquad (3.1)$$

where $\varepsilon_t \sim N(0,\sigma_\varepsilon^2)$ is considered the measurement error. The time variation of the variance of the excess return is assumed to follow a GARCH (1,1) process given by,

$$\sigma_t^2 = \alpha_0 + \alpha_1 \varepsilon_{t-1}^2 + \beta_1 \sigma_{t-1}^2 . \qquad (3.2)$$

The dynamic of the price of risk (b_t) is assumed to follow a simple random walk process described by,

$$b_t = b_{t-1} + v_t \qquad (3.3)$$

where $v_t \sim N(0,\sigma_v^2)$ is uncorrelated with the measurement error.

The only observable quantity in this setup is the excess return from the market. The unknown parameters of the model can by summarised by $\Theta \equiv \begin{bmatrix} \alpha_0 & \alpha_1 & \beta_1 & \sigma_\varepsilon^2 & \sigma_v^2 \end{bmatrix}'$. We can estimate these parameters by recognising that the model is already in the linear state-space framework.

Equation (3.1) is the measurement equation and equation (3.3) gives the state dynamic. The first step in the estimation is to construct the prediction error decomposition form of the likelihood function by applying the Kalman filter. Once this likelihood function is obtained, it can be directly maximised by the numerical method. This gives us the required parameters, as well as the standard error of the estimates. The Kalman filter also generates the most likely estimate of the states, i.e., b_t. This is the quantity which interests us in our further analysis.

The model applied here has an added level of complexity which sets it apart from other linear state-space model estimations. The complexity arises due to the time-varying conditional variance. Chapter 6 of Kim and Nelson (1999) clearly explains the difficulty entailed in this and proposes a workable solution (we see no need to repeat the methodology here). Interested readers will find it fruitful to explore this excellent reference together with the algorithms for Linear Kalman Filter in Chapter 1.

3.1.2 Data Used for Empirical Study

We apply our model to the monthly data on G7 countries (Canada, France, Germany, Italy, Japan, the UK, and the USA) from January 1971 to September 2004. The stock price index for each country is obtained from the Morgan Stanley Capital International Index. The short-term interest rate for each country is obtained from the International Financial Statistics of the International Monetary Fund (IMF). The excess return is calculated in the usual way by subtracting the short-term interest rate from the equity return.

Table 3.1 presents the summary statistics describing the excess return series for each country. The mean excess returns are positive, which reflects the positive performance of the major stock markets during the past three decades. The considerable size of the minima and maxima relative to the mean and standard deviation of the series may indicate that the series are heavy tailed. This possibility is confirmed by a rejection of the normality assumption for the series. The JB-test statistic is computed using skewness and kurtosis parameters.

Table 3.1. Summary Statistics for Excess Return.

	CAN	FRA	GER	ITA	JAP	UKG	USA
Mean	0.0067	0.0245	0.0205	0.0067	0.0274	0.0087	0.0190
Std. Dev.	0.6006	0.7380	0.6935	0.8630	0.6382	0.7291	0.5401
Maximum	1.8719	2.6151	2.3347	3.0967	2.3406	6.3208	2.0038
Minimum	-2.6927	-2.7193	-3.0246	-2.3607	-2.4332	-3.2291	-2.6394
JB-test	82.8944	15.1854	83.0803	25.4032	22.0581	3833.4390	59.4152
P-value	0.0000	0.0005	0.0000	0.0000	0.0000	0.0000	0.0000

JB is the Jarque-Bera test statistic for normality.

3.1.3 Discussion of Empirical Results

Tables 3.2 and 3.3 show the empirical results of the model and the model diagnostics. Table 3.2 reports the maximum likelihood estimates of the parameters of models (3.1), (3.2), and (3.3). Table 3.3 reports the p-value of the Q statistic and Q^2 statistic. The Q and Q^2 tests check for no serial correlation in the standardized and squared standardized residuals up to lag 24, respectively. These results indicate that the null hypothesis of no serial correlation in the standardized and squared standardized residuals is accepted for all countries at a significance level of 0.05 or 0.01. The empirical specification of the model in Table 3.2 is thus clearly supported.

Figures 3.1a through 3.1g depict the movement of the price of risk (b_t) for each country. The general appearance of the plot for the USA is similar to that reported by Jochum (1999): falling markets are associated with drops in the price of risk, while rising markets are associated with rises in the price of risk. This relation is most conspicuous during the rally and subsequent market correction in the U.S. market in the mid eighties. When market participants perceive the market valuation as expensive, they demand (and receive) increased compensation for holding onto their market portfolios. Later, after a major decrease in the average price level, as in 1973 and 1987, the process reverses. In a market where investors expect a major build-up in value, they demand less compensation for risk. The same holds true in

Japan: the Japanese market plummets at the collapse of the bubble at the beginning of 1990, and then stagnates for more than a decade. This pattern implies that the average investor is mainly concerned with the likelihood that the market will drop and yield a negative return.

We have also compared the market volatility as estimated by the time-varying parameter GARCH (TVGARCH) model and the fixed parameter GARCH model for each country. Chou, Engle, and Kane (1992) obtain an almost perfect fit between the two volatility series. Our findings, like those of Jochum (1999), reveal some discrepancy, but the correlation between the two series still exceeds 95% overall (0.976 for Canada, 0.983 for France, 0.981 for Germany, 0.990 for Italy, 0.968 for Japan, 0.955 for the UK, and 0.976 for the USA).

Table 3.4 shows the results of the variance decomposition. The price of risk varies considerably throughout the sample period. In light of this, it will be of interest to estimate the vector autoregression (VAR) model among the price of risk, excess return, and volatility, then report the variance decomposition based on the estimated model (the price of risk in the VAR is first differenced because of its construction as an I(1) variable. The necessary length for the VAR model is determined by the Akaike information criterion (AIC). The result of the variance decomposition depends on the ordering of the variables in the VAR model. Here, therefore, we use the same ordering adopted by Jochum (1999), i.e., the price of risk, excess return, and volatility. The column headings in the table provide the explanatory variables. The panel below indicates the percentage explained during the period.

In the results for the six-month period in the US given in Table 3.4, 94.97% of the variance in the price of risk is attributed to the variance itself, while 2.14% and 2.90% are explained by changes in excess return and volatility, respectively. Readers may also be interested to note that changes in the price of risk explain more than 60% of the variability of the excess return, while volatility explains only about 1.5%. Ferson and Harvey (1991) find that the price of risk is more important than volatility as an explanatory variable for the movements of excess returns in the USA. Our results confirm that finding in all the other countries as well.

Table 3.5 presents the results of our analysis of the stability of the correlation matrix of the price of risk series for the G7 countries. The result is reported by comparing the correlation matrices computed using a sliding sample size of 5 years of data. Thus, our dataset allows us to examine six non-overlapping samples. These correlation matrices will not necessarily be equal over this sample period if local factors substantially influence the price of risk. If these markets are integrated, however, equality among these correlation matrices is quite likely. Here we find inequality of the correlation matrices in five of the six sub-samples. In the sixth, the sub-sample which includes the market crash of October 1987, we observe a relatively stable correlation matrix. This stability only lasts for a single five-year sub-sample and then deviates again. A closer look at the same episode with sliding sample sizes over different timeframes may reveal different patterns. In the context of portfolio diversification, this type of analysis of the stability of correlation matrices plays an important role when examining the return correlations (e.g., Adjaoute and Danthine, 2004 and Silvapulle and Granger, 2001). In this paper, however, we confine our examination to a much more fundamental issue, i.e., the influence of the price of equity market risk on the overall return generation process.

Table 3.6 presents the concordance statistic, a measure of how closely the different phases of the price-of-risk series relate to each other among the G7 countries. The use of concordance as a measure of co-movement has advantages over a correlation analysis. The series must be rendered stationary by differencing with the correlation measure, and the results are easily influenced by the presence of single, large-magnitude observations of the type likely to represent single events. The concordance statistic, in contrast, captures the notion that the prices of risk in two markets move together. We see from the table that the Japanese market moves in phase with the markets of France and Germany, but not with that of the USA. We would similarly expect Canada and USA to move together, and the result supports that view. We also notice that the three main European markets — France, Germany, and the UK — move in phase with each other for most of the time. The geographical proximity and relative homogeneity of these markets may explain this common phase in the movements of their prices of risk.

Table 3.2 Maximum Likelihood Estimates of the Parameters.

	CAN	FRA	GER	ITA	JAP	UKG	USA
α_0	0.0340	0.1286	0.0158	0.1987	0.0052	0.0324	0.0247
	(0.027)	(0.052)	(0.010)	(0.062)	(0.002)	(0.018)	(0.014)
α_1	0.1096	0.1286	0.1489	0.1536	0.0880	0.2937	0.0720
	(0.054)	(0.056)	(0.042)	(0.065)	(0.025)	(0.094)	(0.027)
β_1	0.7983	0.6385	0.8277	0.5870	0.9084	0.7050	0.8435
	(0.116)	(0.112)	(0.047)	(0.103)	(0.024)	(0.093)	(0.057)
σ_ε^2	1.0E-05	8.2E-06	9.8E-07	1.8E-05	3.9E-04	1.0E-08	9.8E-07
	(5.3E-07)	(2.7E-07)	(2.9E-07)	(3.0E-04)	(2.3E-05)	(5.0E-10)	(5.5E-08)
σ_v^2	9.9E-11	1.0E-10	9.9E-11	9.9E-11	1.0E-10	9.9E-11	1.4E-04
	(5.3E-12)	(4.7E-12)	(5.0E-12)	(4.4E-12)	(5.1E-12)	(5.0E-12)	(3.6E-04)

Numbers in parentheses below the parameters are standard errors computed using the Hessian matrix at the point of convergence.

Table 3.3 Model Diagnostics.

	CAN	FRA	GER	ITA	JAP	UKG	USA
Q-Stat	0.142	0.295	0.286	0.100	0.692	0.013	0.748
Q^2-Stat	0.424	0.023	0.559	0.106	0.541	0.040	0.245

The table entries represent p-values for the corresponding tests. The Q-statistic and Q^2-statistic tests check for no serial correlations in the standardized and squared standardized residuals up to lag 24, respectively.

Table 3.4 Decomposition of Variance: Six-Month Forecast Period.

| | | Explanatory Variables | | |
		Price of Risk	Excess Return	Volatility
Canada	Price of Risk	96.83	2.22	0.95
	Excess Return	47.75	49.37	2.88
	Volatility	5.83	6.02	88.15
France	Price of Risk	96.21	0.97	2.82
	Excess Return	47.57	48.80	3.63
	Volatility	1.19	8.79	90.02
Germany	Price of Risk	99.93	0.04	0.03
	Excess Return	46.87	53.08	0.05
	Volatility	1.83	4.18	93.98
Italy	Price of Risk	99.78	0.04	0.18
	Excess Return	42.17	57.79	0.04
	Volatility	0.92	8.12	90.96
Japan	Price of Risk	96.49	2.78	0.73
	Excess Return	65.47	34.37	0.16
	Volatility	0.60	0.53	98.87
U.K.	Price of Risk	92.22	5.82	1.96
	Excess Return	86.43	12.23	1.34
	Volatility	7.07	28.84	64.09
U.S.A.	Price of Risk	94.97	2.14	2.90
	Excess Return	62.99	35.58	1.43
	Volatility	4.39	8.53	87.09

The VAR estimation is based on the following variables and ordering: price of risk, excess return, and volatility, where the price of risk is first differenced. The column headings show the explanatory variables.

Table 3.5 Testing Equality of Correlation Matrices of Price of Risk From Successive Samples.

First Sample Period	Second Sample Period	Statistic
Jan. 1972 – Dec. 1976	Jan. 1977 – Dec. 1981	0.000
Jan. 1977 – Dec. 1981	Jan. 1982 – Dec. 1986	0.000
Jan. 1982 – Dec. 1986	Jan. 1987 – Dec. 1991	0.000
Jan. 1987 – Dec. 1991	Jan. 1992 – Dec. 1996	0.075
Jan. 1992 – Dec. 1996	Jan. 1997 – Dec. 2001	0.001
Jan. 1997 – Dec. 2001	Jan. 2002 – Sept. 2004	0.000

This test statistic has a chi-squared distribution in which the degrees of freedom depend on the number of series used. There are 7 prices of risk series in this study. The entries in the last column represent p-values for the null hypothesis that the two correlation matrices are equal. The brief description of this statistic following from Jenrich (1970) is given separately as an appendix.

Table 3.6 Concordance Statistics between Probability of High Variance State (Full Sample).

	Canada	France	Germany	Japan	Italy	U.K.	U.S.A.
Canada	1	0.5165	0.5063	0.4733	0.5496^{***}	0.4936	0.6768^{*}
France		1	0.9542^{*}	0.8855^{*}	0.5139	0.9109^{*}	0.3715
Germany			1	0.8651^{*}	0.5038	0.9313^{*}	0.3817
Japan				1	0.5063	0.8524^{*}	0.3181
Italy					1	0.5522^{***}	0.6692^{*}
U.K.						1	0.3842
U.S.A.							1

The critical values are computed by the method of McDermott and Scott (1999) on the assumption that the data on the price of risk are generated by random walk without drift. These values are 0.5430, 0.5600 and 0.5896 for 10% (***), 5% (**) and 1% (*) significance levels.

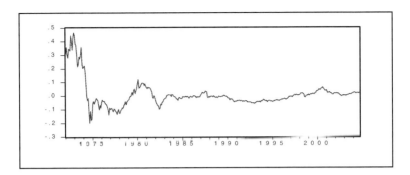

Figure 3.1a Price of Risk for Canada.

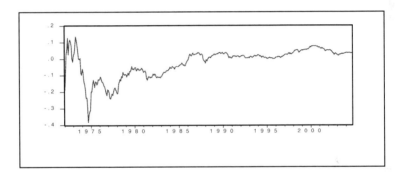

Figure 3.1b Price of Risk for France.

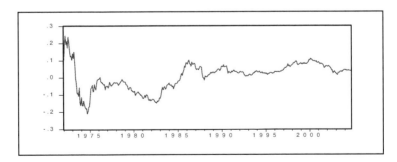

Figure 3.1c Price of Risk for Germany.

Figure 3.1d Price of Risk for Italy.

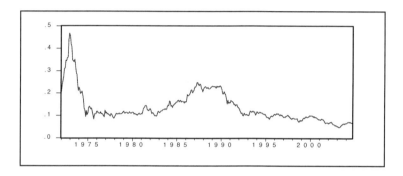

Figure 3.1e Price of Risk for Japan.

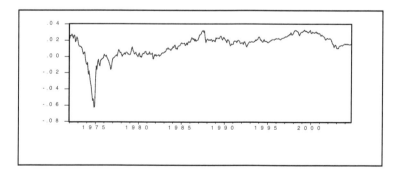

Figure 3.1f Price of Risk for the UK.

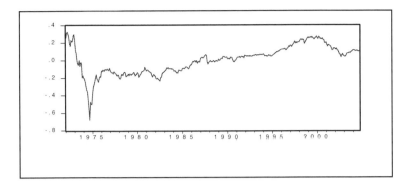

Figure 3.1g Price of Risk for the USA.

3.1.4 Summary of Results

This section analysed the behaviour of the risk premium of the G7 countries. The risk premium is defined as the product of two components, i.e., the amount of risk and the unit price of risk. The price of risk has traditionally been assumed to remain constant over time. If equity markets are financially integrated, the price of risk should be the same across markets. The price of risk may differ across markets, however, if barriers to capital flows across markets or other influences impede the financial integration of markets. In light of this, we decided to investigate the price-of-risk behaviour in several economies. Using the method of Jochum (1999), we began by explicitly considering the variation in the price of risk in the model. The empirical results confirm that the price of risk changes considerably over time for all G7 countries. Our results also tend to indicate that market downturns are associated with drops in the price of risk, whereas rising markets push the price of risk higher. Further, we find that the price of risk is more important than volatility as an explanatory variable for the movements of excess returns.

Our next step was to explore the co-movement of the price of risk in G7 countries. First, we analysed the stability of the correlation matrix of the price-of-risk series for the G7 countries. We found that five of the six sub-samples show inequality of the correlation matrices, whereas the sub-sample which includes the market crash of October 1987 shows a relatively stable correlation matrix. This stability only lasts for a single five-year sub-sample and then deviates again.

Lastly, we used the concordance statistic to capture how closely the different phases of the price-of-risk series relate to each other among the G7 countries. The empirical results indicate that the USA and Canada remain in phase with each other for most of the time. Similarly, the three main European markets also usually move in phase. The geographical proximity, relative homogeneity, and interdependence of these economies may explain this common phase in the movements of their prices of risk.

3.2 Economic Convergence in a Filtering Framework

The introduction of the Euro in January 1999, suggests that a similar currency union could be introduced to other region of the world. Mundell (1961) argued that a currency union can facilitate international trade, and a single medium of exchange reduces transaction costs in regional trade. The new regime would thereby stimulate capital flows and investment, increase growth and employment, and improve the balance of payments performance. Madhur (2002) emphasized that a single currency promotes greater trade among countries; therefore, openness and the volume of intra-regional trade will be greater under a common currency than under a regime of national currencies with floating exchange rates.

At the very heart of the currency union is the issue of economic convergence, which depends on a high degree of convergence of national economies. Economic convergence occurs when poor countries, with low initial income and productivity levels, tend to have high rates of economic growth that would enable them to catch up, or converge upon, the living standards of rich countries over the long run (Lim and McAleer, 2003). This effort can be achieved only when all member nations face both external and domestic shocks in a similar pattern (Jayaraman, Ward, and Xu, 2005). Furthermore, if the responses of national economies to the macroeconomic shocks were asymmetric, the cost of premature monetary integration would be high. Therefore, economic convergence of national economies and symmetry of underlying macroeconomic shocks are the two preconditions for a currency union.

Besides currency union, tests of the convergence hypothesis of neoclassical theory (Solow, 1956), have been a central issue in the empirical studies of growth theory in recent years. These include cross-sectional studies, such as Dowrick and Nguyen (1989), Barro (1991), Barro and Salai-Martin (1994), panel data analysis, Quah (1993) and more recently time-series based tests by Bernard (1992) and Bernard and Durlauf (1995). Time-series studies, which focus on the long run behavior of differences in per capita output across countries, have relied on unit root and co-integration techniques to test for the existence or nonexistence of convergence among a group of countries. Using this framework time-series tests generally reject convergence for a range of countries.

The constant parameter co-integration model, which essentially tests if the linear combination of two series is stationary, is based on the strong assumption that the long-run relationship between two series is time-invariant and linear. Such an assumption is more apt to test for steady state behavior, as pointed out in Bernard and Durlauf (1995). Cross-country differences in incomes, however, suggest that it is more likely that countries have not reached their steady states and may still be in transition. It is also equally likely that the process of convergence itself is nonlinear and subject to structural changes e.g. due to technology and policy shocks. Thus the assumption of 'structural stability' would lead one to reject convergence under such situations. Furthermore, rejection of convergence in these studies, does not tell us whether this occurs because per capita incomes of countries in the sample are diverging from one another or they are in the process of converging, without reaching the limiting state. The difference in the two outcomes would result in significantly different steady state results.

It might seem that time series tests contradict cross-sectional ones, the former not allowing the researcher to decide in favor of convergence. Considering previous results related in St. Aubyn (1995) and Hall and St. Aubyn (1995), it becomes likely that one of the following two assertions is true. First, industrialized countries are converging but co-integration-type tests fail to recognize that because they have low power, especially when structural breaks are present. Second, only some of the considered countries are converging. This would be enough for cross-sectional initial value regressions to display a significant negative coefficient for the initial value, and would further explain the failure of a number of pair-wise co-integration tests.

Given the background above for testing economic convergence, we need a methodology that is robust to structural breaks in the time series as well as that is likely to reveal the stochastic nature of the converging process. Before discussing that we need an acceptable definition of the convergence process.

3.2.1 Defining Convergence

The meaning of convergence' may be clarified before we proceed into testing for it. We are considering two economic series of interest to us. For example, these could be GDPs per head of any two countries. These two are said to converge to each other if their difference becomes stationary in the long term. In more formal terms, consider two economic series X_t and Y_t. These two may represent GDPs per head or any other two series we are interested in. These two series converge if:

$$\left(X_t - Y_t\right) \xrightarrow{\ P\ } \varepsilon_t \text{ as } t \to \infty \qquad (3.4)$$

where ε_t is a random variable obeying the following conditions:

$$E[\varepsilon_t] = D_{XY}, \ \text{Var}(\varepsilon_t) = \sigma^2 < \infty . \qquad (3.5)$$

Equations (3.4) and (3.5) mean that the difference between the two series converges in probability to a third series that is stationary, having a constant mean D_{XY} and a finite constant variance σ^2.

To turn this definition into one compatible with the definitions presented elsewhere (e.g. Barro and Sala-i-Martin (1995)), we could define economic convergence as:

a) Point wise, if $\sigma^2 = 0$;

b) Unconditional, if $D_{XY} = 0$;

c) Conditional, if $D_{XY} \neq 0$.

Thus, the above definitions encompass the "beta-convergence" concept proposed by Barro and Sala-i-Martin (1995). In their study, series are supposed to converge to their steady-state level at an annual constant rate. If steady-states are the same, "beta-convergence" is unconditional. If they are different but grow at the same rate, "beta-convergence" is conditional. Definitions presented in (3.4) and (3.5) are more general because they do not imply a constant rate of convergence and therefore differences between series are not necessarily stationary from the beginning.

Bernard and Durlauf (1995) propose a related definition: series X_t and Y_t converge if the long run forecasts for both series are equal at time t:

$$\text{Lim}_{k \to \infty} \left(X_{t+k} - Y_{t+k} \big| I_t \right) = 0 \qquad (3.6)$$

where I_t represents information at time t.

In some sense, the definition in (3.6) is more stringent: Bernard and Durlauf actually allows for the difference between series to be a random walk. The definition in (3.4) and (3.5) prevents this possibility by imposing a restriction on the variance.

3.2.2 Testing for Convergence

A number of different methods have been proposed to test and measure convergence of economic series, particularly GDPs per head. Here, the main conclusions that arise from some previous work are covered. Some methods use cross-section information when testing for convergence of a group of more than two series. Examples are the analysis of the time series of a dispersion measure, "initial value" or "Barro" regressions and their derivations, random fields regressions and the estimation of Markov transition matrices (St. Aubyn (1999)).

These methods suffer from a common drawback: it becomes difficult to accommodate the possibility of limited convergence. By limited convergence it is understood that only a subset of series (e.g. countries' (GDPs) converge to one leading series. When there is more than one such grouping, there is more than one leading series and each of the remaining series is attracted by one and only one of them.

The methods proposed by Quah (1996) and by Durlauf and Johnson (1992) are specially designed to address this issue. But, their effectiveness relies on strong correlation within the groupings. St. Aubyn (1995) demonstrates this using Monte Carlo method.

Pure time series tests for convergence usually take the difference between two series and analyze its properties. According to St. Aubyn (1999), they share a common characteristic: they make it possible to allow for convergence clubs or limited convergence. For example, if you suppose that country A belongs to the same club as country B, simply test for convergence of series A to B.

The more popular time series approach to convergence is to test this difference for stationarity using, for example, a Dickey-Fuller statistic. This approach is not completely satisfactory, in the sense that convergence, according to the definitions from the previous section, does not necessarily imply stationarity of the differences. For example, Hall, Roberston and Wickens (1992) give the following example: two series are completely independent from each other, and, from a certain point in time, become exactly equal. These series would converge, but would not pass the stationarity test.

In order to allow for a convergence process that starts occurring after the beginning of the period being considered, Hall, Roberston and Wickens (1992) propose the use of the Kalman Filter. St. Aubyn (1999) reports evidence that this method is more powerful than the Dickey-Fuller one when there is a structural break in the convergence process.

3.2.3 Testing Convergence – Dickey-Fuller

Dickey-Fuller convergence tests are stationarity tests for the differences between the two economic series. Equation (3.7) can be estimated using OLS and d_t is the difference between the two series being discussed.

$$d_t - d_{t-1} = b_1 + b_2 d_{t-1} + \sum_{j=1}^{l} b_{2+j} \left(d_{t-j} - d_{t-j-1} \right) + \varepsilon_t . \qquad (3.7)$$

The number of lags denoted by is chosen so that the residuals would not display significant autocorrelation or non normality. The standard tests for autocorrelation and normality could be implemented for this purpose.

The statistic of concern is the t-statistic for the b_2 coefficient. Under the null hypothesis of non-stationary, this statistic is not normally distributed. The relevant critical values are available from MacKinnon (1991).

3.2.4 Testing Convergence – Kalman Filter

Here is the brief description of the method and the detailed description can be found in St. Aubyn (1995). In this approach the difference between the two series, d_t, is modeled as,

$$d_t = \alpha_t + \varepsilon_t, \ \varepsilon_t \sim N\left(0, \sigma_\varepsilon^2\right) \tag{3.8}$$

$$\alpha_t = \alpha_{t-1} + \eta_t, \ \eta_t \sim N\left(0, \Omega_t^2\right), \ \Omega_t^2 = \phi\Omega_{t-1}^2 \tag{3.9}$$

and Ω_0^2 is assumed initial value of the variance of α_t.

The difference between the two series is modeled as a random walk plus noise. Although the noise variance σ_ε^2 is constant through time, the variance of η_t is potentially time varying and given by Ω_t^2. If $|\phi|$ is less than 1, this variance tends to 0 in the long run, meaning that the two series are converging. Their difference becoming a stationary variable: the random walk component vanishes eventually and the noise becomes the only stochastic disturbance.

The model in (3.8) and (3.9) is written in state-space form, equation where the first one is the measurement equation and the second one is the state equation. The likelihood function can be constructed using the Kalman Filter as discussed in Chapter 1. The maximum likelihood estimates for the parameters are obtained by maximizing it.

The null hypothesis of no convergence implies that $\phi = 1$. Therefore, the proper test for convergence is: H0: $\phi = 1$ against H1: $\phi < 1$. The implementation of this test requires the distribution of ϕ under the null. St. Aubyn (1995) shows via simulation the critical values that are applicable for this hypothesis test. The test is based on the quantity,

$$t\left(\hat{\phi}\right) = \frac{\hat{\phi} - 1}{\sqrt{h_\phi^{-1}}} \tag{3.10}$$

where $\hat{\phi}$ is the estimated maximum likelihood value of ϕ and h_{ϕ}^{-1} is the inverse of the diagonal element of the Hessian matrix corresponding to ϕ.

The test described above using Kalman Filter can be easily extended to accommodate differences in steady state values reached by the economic time series being investigated. Finally, the hypothesis that there are differences in the convergence process across the candidate series is addressed. That is to say that even if all series are converging towards the same attracting economy, it may well be the case that they are doing it at different speeds and towards different steady-states.

Consider the following set up,

$$d_t = d_{SS} + b_t \left(d_{t-1} - d_{SS} \right) + \varepsilon_t \tag{3.11}$$

$$b_t - \overline{b} = a_1 \left(b_{t-1} - \overline{b} \right) + \eta_t, \ -1 < a_1 < 1. \tag{3.12}$$

Here, ε_t and η_t are i.i.d. Normal distribution.

b_t is a time-varying speed of convergence. Since a_1 is bounded within the unit circle, b_t tends to return to its "long run" value, given by $\overline{b}$. This is a case of a "return to normality" model. If b_t is smaller than one, the difference between the two series converges to the steady-state value given by d_{SS}.

Equations (3.11) and (3.12) express the measurement and transition equations in a state-space representation of the return to normality model. The concentrated log-likelihood can be written, as discussed in Chapter 1, as an implicit function of the relevant parameters and maximized using numerical methods.

As an illustration of this Kalman Filter based approach to economic convergence, we estimated the models described above with the price of risk series computed as discussed earlier for the UK and the USA economies. We used monthly data covering the period April 1984 to May 2009. In fact we are investigating whether these two economies were converging with respect to the equity price of risk over the sample period.

We compute the difference variable, d_t as the price of risk for USA minus the price of risk for the UK. The statistic based on equation (3.10) is -41.58 indicating the rejection of the no convergence hypothesis at the

1% level. In other words, these two economies represented convergence
of equity price of risk over the sample period.

In order to understand the speed of convergence as well as the steady
state values we also estimated the model given by (3.11) and (3.12). The
estimated parameters with t-statistics in parentheses are given below:

d_{SS}	$\bar{b}$	a_1
1.69	0.89	0.80
(63.64)	(15.72)	(12.57)

These estimated values indicate that the speed of convergence of the
two equity price of risk series was moving to the steady state value
rapidly indicating economic convergence. The positive steady state value
of d_{SS}, shows that the US equity market price of risk remained over that
of the UK.

3.3 Ex-Ante Equity Risk Premium

In this section we discuss a model for the aggregate stock market
together with its dividend yield and earnings yield so that the ex-ante risk
premium could be extracted in an unobserved component modeling
framework. We posit the model as a linked stochastic differential
equation system and the linking variable is the ex-ante risk premium. By
hypothesizing a realistic dynamic structure for the ex-ante risk premium,
we demonstrate how such a system could be estimated as a filtering
problem. As a practical demonstration of the methodology we apply the
model to the U.K. stock market data.

3.3.1 Background to Ex Ante Risk Premium

Finance research has traditionally focused on the relationship between
stock price, dividends and earnings in order to find an acceptable
explanatory model for the observed stock prices. The asset pricing theory
predicts that these are related by a quantity that is essentially unobserved
and it is referred to as the ex-ante risk premium. The estimation problem
of this ex-ante risk premium has spawned a vast empirical literature.

In this section we approach the problem of estimating the ex-ante risk premium by modeling the relationship between asset price, dividend yield, earnings yield and the ex-ante risk premium as a linked system of continuous time diffusion processes. In particular we assume that the dividend yield mean-reverts to some fraction of the earnings yield, and that the earnings yield itself mean reverts to a long run constant level. These (or related) assumptions can be found in earlier literature such as Davidson, Okunev and Tahir (1996), Campbell and Shiller (1988a, b), Campbell and Kyle (1993).

For the ex-ante risk premium there is less guidance in the literature as to an appropriate dynamic model. The simplest assumption (and that we employ here) is that it follows a mean-reverting process. This assumption was found to give good results in Bhar, Chiarella and Runggaldier (2004) that developed an approach to extract information about the index equity risk premium from index derivatives data. Here we assume that the volatility of the ex-ante risk premium is constant over the sample period. Later we would point out how this assumption could be relaxed.

The main contribution of this exposition is to explain how such a model could be implemented in practice and we demonstrate the methodology using aggregate market data from the U.K. We also highlight the computational issues at the appropriate places later in the section.

Next, we discuss the stochastic dynamics for the index, the dividend yield, the earnings yield and the ex-ante risk premium. Following that we describe the filtering setup. Finally, we discuss the data and the empirical results for the constant volatility case before summarizing the analysis.

3.3.2 A Model for Ex Ante Risk Premium

Let us use $S(t)$, $r(t)$, $q(t)$, and $\pi(t)$ to denote respectively the index, the risk-free rate, the dividend yield rate and the ex-ante equity market risk premium at time t. We propose for the movements of the index the geometric Brownian motion process,

$$\frac{dS}{S} = \left[r(t) + \pi(t) - q(t) \right] dt + \sigma_S dZ_S \qquad (3.13)$$

where Z_s is a Wiener process and σ_s is the instantaneous volatility of the index return. In this study we treats σ_s as constant, though we acknowledge that future research may need to consider the dynamics of σ_s also as a diffusion process, thus leading to a stochastic variance model. The intuition expressed by equation (3.13) is that the expected capital gain on the index equals the risk free rate plus the ex-ante equity market risk premium minus the continuous dividend yield.

Inspired by the approach of Chiang, Davidson and Okunev (1996), who model earnings and dividends as a linked diffusion system, we model earnings yield (e) and dividend yield according to,

$$dq = \beta_q \left(\gamma e - q \right) dt + \sigma_q dZ_q \tag{3.14}$$

$$de = \beta_e \left(\overline{e} - e \right) dt + \sigma_e dZ_e . \tag{3.15}$$

Equation (3.14) states that the dividend yield is mean-reverting with speed β_q to some fraction of the earnings yield and expresses in yield form the original idea of Lintner (1956) that firms have in mind a target dividend that is some fixed proportion of earnings. When dealing with index returns it seems appropriate to express this relation in yield form.

Equation (3.15) expresses the notion that the earnings yield is mean-reverting to some long run value $\overline{e}$ with speed β_e. For both processes (3.14) and (3.15) we assume independent Wiener processes Z_q and Z_e and the associated volatilities σ_q and σ_e. It seems unlikely that the Wiener processes Z_s, Z_q, and Z_e would be completely independent, so we allow for correlation amongst all of them i.e. we assume the instantaneous correlations between the pairs of variables defined by, ρ_{Sq}, ρ_{Se}, and ρ_{qe}.

Finally, we model the ex-ante risk premium as a mean reverting stochastic system with its own Wiener process. In other words, we specify,

$$d\pi = \beta_\pi \left(\overline{\pi} - \pi \right) dt + \sigma_\pi dZ_\pi . \tag{3.16}$$

Also, we allow the Wiener process driving the risk premium to be independent of other Wiener processes in the model. Equations (3.13)-(3.16), form a linked stochastic dynamic system. In the next section we show how this linked system could be expressed in the state space framework and the Kalman filter may be applied to estimate the unknown parameters of the system. This estimation process would also allow us to infer the conditional mean and the variance of the ex-ante equity risk premium.

3.3.3 Filtering Ex Ante Risk Premium

It will be convenient to express (3.13) in terms of $s = \ln(S)$, i.e.

$$ds = \left(r + \pi - q - 0.5\sigma_s^2\right)dt + \sigma_s dZ_s. \tag{3.17}$$

With this change of variable equations (3.14), (3.15), (3.16) and (3.17) form a system of linear stochastic differential equations, to which it is appropriate to apply the Kalman filter. We treat s, q and e as observed quantities and the ex-ante risk premium as the unobserved state variable of the system. For the Kalman filtering application it is convenient to discretise the system (3.14) – (3.17), using the Euler-Maruyama scheme. The discretisation of the state equation (3.16) yields,

$$\pi_t = \beta_\pi \bar{\pi} \Delta t + \left(1 - \beta_\pi \Delta t\right)\pi_{t-1} + \sigma_\pi \varepsilon_{\pi,t} \sqrt{\Delta t}, \; \varepsilon_{\pi,t} \sim N(0,1). \tag{3.18}$$

Next, we discretise the measurement equations (3.14), (3.15) and (3.17) and write these in vector notation as,

$$\begin{bmatrix} \Delta s_t \\ \Delta q_t \\ \Delta e_t \end{bmatrix} = \begin{bmatrix} r_{t-1} - q_{t-1} - 0.5\sigma_s^2 \\ \beta_q \gamma e_{t-1} - \beta_q q_{t-1} \\ \beta_e \bar{e} - \beta_e e_{t-1} \end{bmatrix} \Delta t + \begin{bmatrix} \Delta t \\ 0 \\ 0 \end{bmatrix} \pi_t + \begin{bmatrix} \varepsilon_{s,t} \\ \varepsilon_{q,t} \\ \varepsilon_{e,t} \end{bmatrix} \tag{3.19}$$

where, $\varepsilon_s, \varepsilon_q$, and ε_e are standard normal variates. The variance-covariance structure of the noise terms in the observation equation (3.19) is given by,

$$\begin{bmatrix} \sigma_s^2 & \rho_{s,q}\sigma_s\sigma_q & \rho_{s,e}\sigma_s\sigma_e \\ \rho_{s,q}\sigma_s\sigma_q & \sigma_q^2 & \rho_{q,e}\sigma_q\sigma_e \\ \rho_{s,e}\sigma_s\sigma_e & \rho_{q,e}\sigma_q\sigma_e & \sigma_e^2 \end{bmatrix} \Delta t. \tag{3.20}$$

The model parameter vector is, therefore,

$$\Theta \equiv \left(\sigma_s, \beta_\pi, \bar{\pi}, \sigma_\pi, \beta_q, \gamma, \sigma_q, \beta_e, \bar{e}, \sigma_e, \rho_{s,q}, \rho_{s,e}, \rho_{q,e}\right).$$

These parameters are to be estimated by maximizing the prediction error form of the likelihood function described in Chapter 1. The important observation to be made is that the measurement equation (3.19) is linear in the state variable (π).

3.3.4 Ex-Ante Risk Premium for UK

We have taken the monthly earnings and dividend yields and index price data for the FTSI for the period February 1973 to February 2003, from Data Stream. For the risk-free rate we the UK 3-month Treasury bill rates.

The parameter estimates are displayed in Table 3.7. The results of various model diagnostic tests are displayed in Tables 3.8. The filtered equity risk premium is displayed in Figures 1. We see from Table 1 that most estimates are significant. However, β_π, the speed of mean reversion of the risk-premium seems an important exception and the estimate for β_e also has relatively large standard error. Table 3.8 indicates a reasonably good model fit using a number of diagnostics.

3.3.5 Summarizing Ex-Ante Risk Premium for UK

We have set up the relationship between the stock market index level, the dividend yield, the earnings yield and the ex-ante risk premium as a system of stochastic differential equations. We have used unobserved component modeling approach and Kalman filtering methodology to estimate the model and obtain filtered estimates of the ex-ante risk premium. The long run levels for the risk premium are consistent with the point estimates obtained from various ex-post regression based studies. The empirical results also suggest that there are at least two different regimes that classify the behavior of the ex-ante risk premium in the U.K over the sample period.

The methodology developed here shows promise and may be extended in different directions further. First, the indication of the regime changes in the unobserved component could be captured in a hidden Markov framework. Second, some practical application of the one-step ahead prediction available from the filter may be explored. For example, the predicted risk premium may be used to restructure stock portfolio to take advantage of possible excess return.

Table 3.7 Parameters for UK Ex-Ante Risk Premium.

Parameters	U.K.
σ_S	2.45369
	(0.08353)
β_π	0.55231
	(0.33519)
$\bar{\pi}$	4.23528
	(0.45011)
σ_π	0.59899
	(0.27487)
β_q	0.14295
	(0.07284)
γ	0.43442
	(0.12988)
σ_q	0.01279
	(0.00043)
β_e	0.02240
	(0.06499)
$\bar{e}$	0.00005
	(0.00140)
σ_e	0.03484
	(0.00116)
ρ_{Sq}	-0.91965
	(0.00735)
ρ_{Se}	-0.82044
	(0.01541)
ρ_{qe}	0.92970
	(0.00649)

Standard errors are in parentheses below the parameter estimates.

Table 3.8 Residual Diagnostics and Model Adequacy Tests.

Equations	Portmanteau	ARCH	KS Test	MNR
ΔS	0.176	0.682	0.176	0.980
Δ Div Yield	0.020	0.068	0.141	0.996
Δ Earn Yield	0.001	0.035	0.193	0.999

Entries are p-values for the respective statistics except for the KS statistic. These diagnostics are computed from the recursive residual of the measurement equation, which corresponds to the spot index process. The null hypothesis in portmanteau test is that the residuals are serially uncorrelated. The ARCH test checks for no serial correlations in the squared residual up to lag 26. Both these test are applicable to recursive residuals as explained in Wells (1996, page 27). MNR is the modified Von Neumann ratio test using recursive residual for model adequacy (see Harvey (1990, chapter 5). KS statistic represents the Kolmogorov-Smirnov test statistic for normality. 95% significance level in this test is 0.072. When KS statistic is less than 0.072 the null hypothesis of normality cannot be rejected at the indicated level of significance.

3.4 Concluding Remarks

This chapter has demonstrated different applications of the filtering approaches in modeling aspects of equity market. First, the notion of price of risk was introduced and the appropriate model was developed to extract information of unobserved time varying price of risk. Second, a model was developed to show how to investigate time varying process of economic convergence. Finally, a linked system of stochastic processes was developed to infer ex-ante equity market risk premium.

Chapter 4

Filtering Application – Inflation and the Macroeconomy

This chapter shows how filtering application may be useful in understanding complex dynamics in understanding the link between inflation uncertainty and interest rates. It becomes more interesting when this is set against the background for five inflation targeting countries i.e. Canada, Finland, Spain, Sweden and the United Kingdom. The innovation here is to be able to model and extract two different components of uncertainty in a macroeconomic time series.

The sample period is divided into two sub-periods: before and after the inflation targeting period. The inflation uncertainty is decomposed into two components – impulse and structural using a structural time series framework. The results show mixed experiences among the countries. In general, there is a positive association between the expected inflation and interest rates for Canada, Finland, Spain and Sweden for the full period. Expected inflation also has positive and significant association for Finland. Structural uncertainty has positive and significant effect on interest rates for the full period for Spain. Other results are mixed and some not significant contrary to what would have been expected based on extant macroeconomic literature.

4.1 Background and Macroeconomic Issues

During the early 1990's several countries adopted explicit inflation targeting (IT) as a tool for monetary policy under the operational independence of the Central Bank. They recognized the benefits of price stability and consequently adopted it as the principle goal of monetary policy. The effect of inflation on economic performance is an important but complex topic, because it may influence growth negatively. Mallik and Chowdhury (2002) and Cameron, Hum and Simpson (1996) study the inflation-growth relationship. In a recent cross-country study Bruno and Easterly (1998) conclude that the ratio of fervent beliefs (that inflation is harmful to growth) to tangible evidence seems unusually high.

While cross-country studies are affected by extreme values, Friedman (1973) points out, "Historically, all possible combinations have occurred: inflation with and without development (economic growth), no inflation with and without development." Other than actual inflation the inflation uncertainty may also effect the economic growth either way. Therefore it is also important to study the effect of inflation uncertainty on inflation. Grier and Grier (2006), Nas and Perry (2000), Johnson (2002), and Kontonikas (2004) are some of the relevant studies in this context.

Friedman (1977) in his Nobel address claims that inflation and nominal uncertainty positively correlated and causation runs from inflation to uncertainty about future inflation. An increase in inflation may induce an erratic policy response by monetary authority which creates more uncertainty about future rate of inflation. Demetriadas (1988) shows the asymmetric information responsible in this causation, but did not mention about the direction of causality. Increased inflation uncertainty may increase unanticipated inflation and hence will be associated with the costs of anticipated and nominal uncertainty effects interest rates, Huizina (1993).

Wilson (2006) studied the link between inflation, inflation uncertainty and output growth using EGARCH-M (Exponential Generalized Autoregressive Heteroscedasticity in Mean) model and found that increased inflation uncertainty increase inflation and lower economic growth for Japan. Wilson and Culver (1999), Grier and Perry (2000),

Hayford (2000), Fountas et al. (2002) and Grier et al. (2004) found a similar negative relationship between inflation uncertainty and output growth, while, Levin and Renelt (1992), Levin and Zervos (1993) and Clark (1997) failed to provide evidence of any such relationship.

The aim in the study of this chapter is to explore the relationship between inflation uncertainty and interest rates. Interest rates and price are important variables in the macro-economy and are often monitored by the policymakers, especially for the IT purposes. The relationship between these variables has been subject to substantial research. Wilcox (1983), Uribe (2002), Berument and Jelassi (2002), Fahmy and Kandil (2002) and Kandil (2005) focused on the relationship between prices and interest rates. In general, inflation uncertainty affects the economy by increasing long term interest rates. Ball and Tchaidze (2002) on page 108 said that "A large literature argues that monetary policy under Alan Greenspan is well explained by simple reaction function. Interest rates rise when inflation rises and fall when there is a greater economic risk."

The relationship between interest rates and inflation risk is first mentioned by Fisher. Fishers Hypothesis suggests that interest rates and inflation (or expected inflation) are positively related to each other. There are different opinions regarding this type of relationship. Boudoukh and Richardson (1993) argue that this positive relationship exists at all horizon length. Mishkin (1992) found a long run relationship for USA in the long run but failed to establish the Fisher relationship in the short run. Yuhn (1996) found stronger Fisher effect over long horizons for Germany. Berument et al. (2005) studied the relationship between three different types of inflation uncertainty with the interest rates of the UK before and after the inflation targeting period, and supports the notion of inflation targeting regimes. In this analysis we extended the above research from UK to other inflation targeting countries to examine the influence of inflation uncertainty on interest rates before and after the inflation targeting regime.

4.2 Inflation Targeting Countries and Data Requirement

New Zealand was the first country to formally adopt an inflation target of 0-2% in March 1990.

Countries to follow New Zealand in formally adopting inflation-targeting include Canada (1-3%, adopted in February 1991 aimed to achieve by 1995), the UK (1.5-3.5%, adopted in October 1992 aimed to achieve by 1997), Australia (2-3%, adopted on March 1993), Sweden (1-3%, adopted in 1993 aimed to achieve by 1995), Finland (2%, adopted in 1993) and Spain (less than 3%, adopted in 1994). All these countries have achieved an average annual inflation rate of less than 2 per cent, except the UK (2.6%), though it is in its targeting range.

Australia and New Zealand are excluded from the study because monthly data for Consumer Price Index is not available for these countries.

Monthly data has been used for this study. All interest rates and inflation data (change in natural log of Consumer Price Index) were collected from the International Financial Statistics (International Monetary Fund) and the Industrial Production data were collected from Data Stream. Other variables needed for this analysis are monthly interest rates are $\left(R_{mt} \right)$,

$$R_{m,t} = \left[(1 + \tfrac{r_t}{100})^{1/12} - 1 \right] \times 100 \qquad (4.1)$$

where r_t is the annual interest rate, output gap $\left(G_t \right)$,

$$G_t = (\ln Y_t - \ln Y_t^p) \times 100 \qquad (4.2)$$

where Y_t and Y_t^P are industrial production and potential industrial production in period t.

4.3 Model for Inflation Uncertainties

Following Berument et al (2005) we modeled the inflation series with time varying parameters that allow us to extract two forms of inflation uncertainties, i.e. the structural and impulse response components. To be precise, the time series dynamic of the inflation (π_t) is given by the following equations:

$$\pi_t = \phi_{0,t} + \phi_{1,t}\pi_{t-1} + \phi_{2,t}\pi_{t-2} + \varepsilon_t. \tag{4.3}$$

This autoregressive structure using time invariant parameters is found to be adequate using standard statistical tests in Eviews™. The error term ε_t has the GARCH (1, 1) type variance given by:

$$\varepsilon_t \sim N(0, h_t), \quad h_t = \alpha_0 + \alpha_1\varepsilon_{t-1}^2 + \beta_1 h_{t-1}. \tag{4.4}$$

To complete the specification of the inflation time series dynamic, we specified the time varying parameter dynamics as random walk without trend. In matrix notation the state dynamics is given by:

$$\begin{bmatrix} \phi_{0,t} \\ \phi_{1,t} \\ \phi_{2,t} \end{bmatrix} = \begin{bmatrix} 1 & 0 & 0 \\ 0 & 1 & 0 \\ 0 & 0 & 1 \end{bmatrix} \begin{bmatrix} \phi_{0,t-1} \\ \phi_{1,t-1} \\ \phi_{2,t-1} \end{bmatrix} + \begin{bmatrix} \eta_{0,t} \\ \eta_{1,t} \\ \eta_{2,t} \end{bmatrix}. \tag{4.5}$$

The vector of the noise term in the state equation above is assumed to have a normal distribution with a diagonal covariance matrix or in other words these noise terms are assumed to be uncorrelated. This specification is expressed as,

$$\begin{bmatrix} \eta_{0,t} \\ \eta_{1,t} \\ \eta_{2,t} \end{bmatrix} \sim N\left(\begin{bmatrix} 0 \\ 0 \\ 0 \end{bmatrix}, \begin{bmatrix} \sigma_{\eta_0}^2 & 0 & 0 \\ 0 & \sigma_{\eta_1}^2 & 0 \\ 0 & 0 & \sigma_{\eta_2}^2 \end{bmatrix} \right). \tag{4.6}$$

The state dynamic given by equation (4.5) describes the evolution of the time varying parameters of the inflation process leading to the observation or measurement in equation (4.3). In matrix notation this can be expressed as,

$$\pi_t = \begin{bmatrix} 1 & \pi_{t-1} & \pi_{t-2} \end{bmatrix} \begin{bmatrix} \phi_{0,t} \\ \phi_{1,t} \\ \phi_{2,t} \end{bmatrix} + \varepsilon_t . \tag{4.7}$$

The system represented by the equations (4.5) and (4.7) is in state space form and the methodology to estimate the model given the observation on inflation requires application of the Kalman filter described in Chapter 1 in detail. Under the assumption of conditional normal distribution of the error terms the Linear Kalman Filter algorithm may be directly applied.

But the presence of the GARCH error in the measurement equation implies a departure from the main assumptions of the filtering algorithm. The modification necessary to adapt to this situation has been described by Harvey, Ruiz and Sentana (1992) and further insight and illustrations may be found in chapter 6 of Kim and Nelson (1999).

In order to explain the mechanics of separating the inflation uncertainty into two components – impulse $\left(V_t^1 \right)$ and structural $\left(V_t^S \right)$ – we need to refer to the adaptive algorithm of the Kalman filter. This algorithm is has been described in detail in Chapter 1. The enhancement needed to accommodate GARCH type variance is explained here with reference to chapter 6 of Kim and Nelson (1999). To conserve space we simply describe the connection of our model to that reference and point out the parts that we focus on as impulse and structural uncertainties. The chapter 6 and in particular section 6.1 in Kim and Nelson (1999) shows how to implement structural time series model in state space framework with GARCH measurement error. The equation 6.29 in this section is the most important relation that separates the variance in the two components. The first part of the equation 6.29 on the right hand side is the structural component and the second part is the impulse component. Once this connection is understood, the numerical optimization of the likelihood function can be carried out as discussed earlier. The two parts of equation 6.29 needs to be saved at the point where the likelihood function reaches its maximum.

We implement this algorithm in Gauss™ and estimate the model parameters. There are six unknown parameters in this model and these are, $\Theta \equiv \begin{bmatrix} \alpha_0 & \alpha_1 & \beta_1 & \sigma_{\eta_0}^2 & \sigma_{\eta_1}^2 & \sigma_{\eta_2}^2 \end{bmatrix}$. The algorithm allows us to get the filtered estimate of the elements of the state vector which are the three time varying parameters in equation (4.3).

4.4 Testing Fisher Hypothesis

Once the two components of inflation uncertainties are estimated, Fisher hypothesis may be examined with the following basic model:

$$R_{m,t} = \delta_0 + \gamma_1 \pi_t^E + e_t . \qquad (4.8)$$

In this equation, strong form of Fisher hypothesis implies $\gamma_1 = 1$ and the weak form of the hypothesis requires $0 < \gamma_1 < 1$.

In order to incorporate the short run dynamics and to avoid the problem of misspecification, which manifests autocorrelation the right hand side of equation (4.8) can be written as:

$$R_{m,t} = \delta_0 + \gamma_1 \pi_t^E + \sum_{i=1}^{n} \delta_i R_{m,t-i} + v_t . \qquad (4.9)$$

We have further extended the model and incorporated the structural and impulse uncertainties along with the steady state uncertainty and output gap, which may capture the credibility of central banks in the longer term to control inflation.

The interest rate $R_{m,t}$ specification is given by the following equation:

$$R_{m,t} = \delta_0 + \gamma_1 \pi_t^E + \gamma_2 G_t + \gamma_3 V_t^I + \gamma_4 V_t^S + \sum_{i=1}^{n} \delta_i R_{m,t-i} + v_t . \qquad (4.10)$$

In the above interest rate equation π_t^E refers to the expected inflation as captured by the time series model described above and is obtained from the Kalman filter recursive algorithm. The output gap denoted by G_t is the difference between the log output and its trend value obtained by Hodrick-Prescott filter using EViews.

4.5 Empirical Results and Analysis

Table 4.1 shows the mean and standard deviation of economic growth, interest rates, inflation and expected inflation for five countries under study for the full period, before and after the inflation targeting period. It is clear from the table that after the inflation target period, the growth rate for all the countries increased, inflation has reduced to within the targeted range expect for Spain. The average inflation for Spain is 3.17% after the inflation targeting – a shortfall of 0.17% less than the actual target. Interest rates have also decreased for all the countries after the inflation targeting period. Moreover for all variables the volatility is decreased after the inflation targeting period. Therefore, from the visual description it looks like that the inflation targeting is somehow successful.

Table 4.2 gives the parameter estimates of the inflation model. The impulse uncertainty for the inflation dynamic is captured by the parameters α_0, α_1 and β_1. For all the series most of these parameters are statistically significant. This component of inflation uncertainty represents the shocks that hit the economy. In a GARCH specification ($\alpha_1 + \beta_1$) denotes the persistence of the shocks. In this sense, the persistence of such shocks is highest in Spain and lowest for Canada. For the other three countries this is very similar. In general, the persistence of impulse uncertainty is high for all the series.

The structural parameters of the inflation dynamic, represented by the parameters ϕ_1 and ϕ_2, capture the changes in association with past inflation to the present realization, and ϕ_0 is indicative of the present level of the inflation. Since these are all time varying the uncertainty introduced by this time variation is the structural component of inflation uncertainty. In other words the time varying parameters show how the shocks hitting the economy propagate through the system. The structural time series model of inflation implemented in this study allows us to separate these two components easily and examine any differing behavior subsequently. In the case of Finland, the variance of the noise term of one of the autoregressive parameters is significant. However, the level component captured by ϕ_0 has the variance of the noise term significant for all the series.

The overall indication from this result is that the structural component of inflation uncertainty is important for all the samples examined in this study, but the behavior differs from country to country. It is also interesting to visualize the patterns of structural inflation uncertainty from Figure 4.1. This plots the sum of the two auto-regressive parameters of the inflation equation (4.3), $(\phi_1 + \phi_2)$. When the estimated sum is negative it would indicate some error correction mechanism at work. In this sense, Canada, Finland, Spain and Sweden exhibit this feature over much of the sample period examined here. In other words, as inflation tends to increase, the monetary authorities take corrective action. When this sum is positive it tends to indicate high persistence as in the case of U.K. This may indeed lower the structural uncertainty of inflation.

Table 4.3, 4.4 and 4.5 demonstrate the estimates of the coefficients of the Fisher equation using ordinary least square method for the full period, pre and post inflation targeting period. Table 4.3 shows the estimated coefficients and the t-statistics for equation 4.8 for the full period. We have considered different types of interest rates as dependent variables. Almost for all the countries, the coefficient of the output gap is positive except for the money market rates (MMR) for Spain and long-term interest rates (LTR) for Sweden. But none of the coefficients are significant, which is expected and parallel to the findings of Berument et al. (2005). Expected inflation shows a positive and significant effect on interest rates except for U. K. as expected. Therefore, we can say that the Central Bank increased the interest rates when the expected inflation increased. The positive risk premium is also an important result and agrees with other research results. The estimated coefficients for Impulse and Structural uncertainty are insignificant for most of the cases except for Spain and for LTR for UK. Some of the results are different from Berument et al. (2005), which may be due to the fact that we have considered Industrial Production in lieu of Real GDP for the calculation of the GAP.

The expected inflation and the output gap coefficients for the pre-inflation targeting period are insignificant for most of the countries expect for Canada. Impulse and structural inflation uncertainty has no effect on interest rates for this period.

For the post-inflation targeting period the coefficients are positive and significant for Finland. Other results are similar. If we compare the effect of expected inflation, structural uncertainty, impulse uncertainty and output gap we can not see any significant difference for the three said periods. However, it is clear that for Finland and Sweden the Central bank act on the expected inflation much quicker after the inflation targeting period. The coefficient of first lag of the interest rate is significant and for most of the countries it is more than one.

4.6 Concluding Remarks

The results are rather conflicting regarding the effect of inflation uncertainty on interest rates. But one can reach some interesting conclusion from this study if the observed long run effect of inflation on interest rates is considered. We note that the estimated value of $(1 - \delta_1 - \delta_2 - \delta_3)^{-1}\gamma_1$ is less than one for all the countries under study for the post-inflation targeting period and more than one for Finland (lending rates), Sweden (Bank rates) and for the United Kingdom (Inter-Bank rates) for pre inflation targeting period. The estimated long run coefficient of 1.47% for Sweden suggests that for a 1% increase of the expected inflation, the Central Bank increased the interest rates by more than one percent, or the real interest rates increased by 0.47%.

Therefore, after the inflation targeting period, the real interest rates have decreased for all the countries under study and the respective Central Banks are less aggressive to act on the interest rate increase. It is also clear from Table 4.1 that the volatility of expected inflation was much lower during the post-inflation targeting period and therefore, there is no need to react as promptly. Moreover the inflation is well under control after the inflation targeting period. These results are also important from policy perspective. In general the purpose of the monetary authority is to eliminate the uncertainty arising from higher inflation. It is clear that the monetary authority successfully eliminated these uncertainties and consequently controlled the level of inflation as well as the uncertainties.

Table 4.1 Descriptive Statistics for Different Periods.

	Full Period		Before IT		After IT	
	Avg.	S. Dev.	Avg.	S. Dev.	Avg.	S. Dev.
Canada						
Growth	2.90	0.18	2.77	0.32	3.10	1.20
Interest rates	7.37	0.30	10.05	0.21	4.32	0.13
Inflation	4.40	0.40	6.69	0.40	1.81	0.30
Exp. inflation	4.63	0.27	6.99	0.35	1.95	0.08
Finland						
Growth	2.71	0.52	1.73	0.11	4.02	0.25
Interest rates	4.42	0.10	5.08	0.08	3.42	0.03
Inflation	5.11	0.54	7.58	0.56	1.26	0.29
Exp. inflation	5.65	0.40	8.40	0.35	1.39	0.10
Spain						
Growth	2.77	0.21	2.38	0.22	3.73	0.18
Interest rates	8.98	0.40	12.76	0.21	4.68	0.20
Inflation	7.90	0.69	10.99	0.73	3.17	0.35
Exp. inflation	8.46	0.46	11.82	0.37	3.30	0.13
Sweden						
Growth	2.02	2.77	1.53	2.95	2.72	2.51
Interest rates	6.29	0.25	8.25	0.13	3.19	0.14
Inflation	5.15	0.62	7.73	0.64	1.14	0.37
Exp. inflation	5.73	0.33	8.40	0.22	1.60	0.11
United Kingdom						
Growth	2.46	0.25	2.03	0.30	3.05	0.15
Interest rates	8.23	0.26	10.26	0.19	5.07	0.08
Inflation	6.31	0.68	8.74	0.76	2.54	0.34
Exp. inflation	6.75	0.41	9.41	0.38	2.62	0.07

IT represents Inflation Targeting.

Table 4.2 Parameter Estimates of Inflation Model.

	α_0	α_1	β_1	$\sigma^2_{\eta_0}$	$\sigma^2_{\eta_1}$	$\sigma^2_{\eta_2}$
Canada	0.0270^{***}	0.4128^{***}	0.3606^{***}	0.0013^{*}	1.59E-07	6.78E-10
	(3.49)	(3.75)	(3.61)	(1.94)	(0.004)	(0.002)
Finland	0.0015	0.0492^{*}	0.9351^{***}	0.0012^{*}	0.0014^{*}	0.0011
	(1.22)	(1.70)	(26.98)	(1.89)	(1.79)	(0.62)
Spain	0.0008	0.1234^{***}	0.8765^{***}	0.0006^{***}	0.0001	0.00013
	(1.23)	(4.27)	(30.36)	(2.50)	(0.50)	(1.44)
Sweden	0.0127^{*}	0.0431^{**}	0.9078^{***}	0.0011^{*}	2.81E-09	6.29E-05
	(1.65)	(1.97)	(19.86)	(1.69)	(0.007)	(0.37)
UK	0.0052^{***}	0.0186^{***}	0.9552^{***}	0.0009^{**}	1.55E-09	1.92E-08
	(2.49)	(5.76)	(98.05)	(1.98)	(0.120)	(0.17)

The numbers in parentheses below the parameter estimates are t-statistics computed using the robust approach suggested in Hamilton (1994), page 145. Data set spans the period February 1973 to January 2006 for all countries.

In Tables 4.2 – 4.5, *, ** and *** indicate significant at 10%, 5% and 1% levels respectively. I.B. refers to Inter Bank rates in Tables 4.3 - 4.5.

Table 4.3 Fisher Equation Estimates (Full Sample).

	Const.	Exp. Inf.	Gap	Inf.(t-1)	Inf.(t-2)	Inf.(t-3)	Str. Var.	Imp. Var.
Canada								
T-bill	0.015	0.053***	0.302***	1.116***	-0.064	-0.015**	-0.068	0.017
	(0.795)	(3.350)	(4.037)	(21.027)	(-0.816)	(-2.202)	(-0.075)	(0.462)
MMR	0.060	0.021***	0.611***	0.701***	0.120*	0.071	-0.427	0.022
	(1.632)	(3.995)	(4.464)	(13.404)	(1.885)	(1.382)	(-0.886)	(0.313)
LTR	-0.002	0.020*	0.015***	1.000***	-0.115	0.092*	0.100	-0.002
	(-0.141)	(2.040)	(3.070)	(14.037)	(-1.561)	(1.752)	(0.640)	(-0.740)
Finland								
LR	0.007	0.014***	0.003	1.103***	0.096	-0.207***	0.002	-0.023
	(1.20)	(2.56)	(0.43)	(20.18)	(1.16)	(-3.80)	(0.07)	(-1.13)
STR	0.019	0.036***	0.013	1.441***	-0.644***	0.170***	-0.163	0.028
	(1.09)	(2.04)	(0.59)	(27.20)	(-7.41)	(3.24)	(1.58)	(0.41)
LTR	-0.007	0.028**	0.009	1.097***	-0.081	-0.042	-0.053	0.046
	(-0.63)	(2.25)	(0.76)	(16.55)	(-0.82)	(-0.66)	(-0.97)	(1.01)
Spain								
T-bill	-0.003	0.005	0.012	0.945***	0.170**	-0.143**	0.189	-0.019
	(-0.30)	(0.25)	(0.54)	(15.29)	(2.01)	(-2.31)	(0.94)	(-0.40)
MMR	-0.024	0.099***	-0.070	0.879***	-0.104	0.074	1.641***	-0.248***
	(-0.82)	(2.74)	(-1.05)	(17.15)	(-1.57)	(1.57)	(4.26)	(-3.78)
LTR	-0.005	0.012*	0.001	1.298***	-0.310	-0.003	0.181**	-0.032
	(-0.88)	(1.78)	(0.11)	(22.43)	(-0.05)	(-0.05)	(1.96)	(-1.48)
Sweden								
BR	0.014***	0.031***	0.013	0.969***	0.012	-0.002	0.021	-0.095***
	(3.07)	(3.44)	(1.16)	(18.52)	(0.17)	(-0.04)	(0.02)	(2.66)
STR	0.032	0.048***	0.000	1.083***	-0.121	0.052	0.049	-0.064
	(1.23)	(3.12)	(0.03)	(19.52)	(-1.61)	(0.98)	(0.28)	(-0.93)
LTR	0.022**	0.022***	-0.003	1.232***	-0.217***	-0.036	-0.031	-0.027
	(2.27)	(3.57)	(-0.46)	(23.66)	(-2.64)	(-0.68)	(-0.49)	(-1.14)
U.K.								
I. B.	0.040	0.041	0.084	0.621***	0.122**	0.217***	0.547	-0.217
	(0.66)	(1.33)	(1.04)	(12.07)	(2.02)	(4.14)	(1.48)	(-1.17)
T-bill	-0.011	-0.044	0.010	1.350***	-0.389***	0.013	0.066	0.037
	(-0.44)	(-0.03)	(0.32)	(25.70)	(-4.53)	(0.25)	(0.45)	(0.50)
STR	-.039*	-0.012	0.005	1.342***	-0.484***	0.101*	-0.133	0.188*
	(-1.88)	(-1.13)	(0.17)	(25.82)	(-5.80)	(1.95)	(-1.03)	(2.75)
LTR	-0.038**	-0.008	0.087	1.314***	-0.530***	0.174***	-0.166*	0.188***
	(2.41)	(-1.00)	(0.35)	(25.61)	(-6.57)	(3.49)	(-1.75)	(3.60)

Table 4.4 Fisher Equation Estimates (Sample Prior to Inflation Targeting).

	Const.	Exp. Inf.	Gap	Inf.(t-1)	Inf.(t-2)	Inf.(t-3)	Str. Var.	Imp. Var.
Canada								
T-bill	0.044	0.055**	0.369***	1.242***	-0.369***	0.045	-0.202	0.053
	(1.33)	(2.14)	(3.47)	(15.79)	(-3.06)	(0.59)	(-0.46)	(0.56)
MMR	0.153**	0.143***	0.692***	0.636***	0.128	0.060	-1.110	0.132
	(2.05)	(2.70)	(3.18)	(8.60)	(1.46)	(0.81)	(-1.15)	(0.63)
LTR	0.019	0.030*	0.076**	0.978***	-0.119	0.095	-0.009	0.010
	(0.72)	(1.93)	(2.55)	(13.36)	(-1.16)	(1.31)	(-0.01)	(0.24)
Finland								
LR	0.029	0.002	0.002	0.992***	0.117	-0.133*	0.035	-0.037
	(1.54)	(0.20)	(0.23)	(12.89)	(1.08)	(-1.72)	(0.72)	(-1.14)
STR	0.057	0.005	0.018	1.304***	-0.497***	0.119*	-0.060	0.054
	(1.58)	(0.18)	(0.58)	(18.92)	(-4.56)	(1.81)	(-0.41)	(0.55)
LTR	0.014	-0.034	0.002	0.993***	-0.042	-0.094	-0.052	0.259
	(0.35)	(-0.18)	(0.10)	(8.75)	(-0.26)	(-0.85)	(-0.47)	(1.64)
Spain								
T-bill	0.010**	-0.045	0.025	0.824***	0.175	-0.135	0.223	0.000
	(2.08)	(-1.12)	(0.66)	(9.27)	(1.47)	(1.49)	(0.67)	(0.01)
MMR	0.129*	0.025	-0.008	0.834***	-0.104	0.033	1.639***	-0.30***
	(1.71)	(0.99)	(-1.00)	(12.17)	(-1.20)	(0.51)	(3.26)	(-2.27)
LTR	0.016	0.013	-0.003	1.296***	-0.314***	-0.018	-0.211*	-0.036
	(0.78)	(1.33)	(-0.20)	(17.71)	(-2.64)	(-0.25)	(-1.75)	(-0.96)
Sweden								
BR	0.048***	0.018	0.012	0.996***	0.024	-0.044	-0.044	-0.065
	(2.73)	(1.47)	(1.36)	(14.53)	(0.24)	(-0.63)	(-0.54)	(-1.58)
STR	0.046	0.018	-0.001	1.101***	-0.227**	0.072	0.032	-0.033
	(1.27)	(0.67)	(-0.07)	(15.59)	(-2.19)	(1.02)	(0.15)	(-0.38)
LTR	0.036**	0.016*	-0.002	1.133***	-0.149	-0.016	0.001	-0.035
	(2.12)	(1.74)	(-0.35)	(16.09)	(-1.41)	(-0.23)	(0.01)	(-1.21)
U.K.								
I.B	0.085	0.017	0.099	0.637***	0.072	0.227***	0.472	-0.202
	(0.84)	(0.34)	(0.79)	(9.04)	(0.83)	(3.16)	(0.97)	(-0.71)
T-bill	0.027	-0.011	0.014	1.309***	-0.355***	-0.011	0.023	0.054
	(0.62)	(-0.54)	(0.27)	(18.29)	(-3.09)	(-0.15)	(0.12)	(0.48)
STR	0.019	-0.018	-0.001	1.274***	-0.422***	0.035	-0.193	0.228**
	(0.54)	(-1.11)	(-0.03)	(17.96)	(-3.80)	(0.49)	(-1.17)	(2.32)
LTR	0.017	0.008	0.003	1.247***	-0.506***	0.124*	-0.162	0.226***
	(0.66)	(0.61)	(0.11)	(17.69)	(-4.79)	(1.85)	(-1.40)	(3.19)

Table 4.5 Fisher Equation Estimates (Sample Post Inflation Targeting).

	Const.	Exp. Inf.	Gap	Inf.(t-1)	Inf.(t-2)	Inf.(t-3)	Str. Var.	Imp. Var.
Canada								
T-bill	0.034	-0.013	0.087	0.876***	0.381***	-0.310***	-0.319	0.062
	(1.24)	(0.37)	(0.76)	(11.67)	(3.92)	(-4.16)	(-0.86)	(1.45)
MMR	0.039	-0.003	0.145	1.250***	-0.374***	0.086	-0.414	-0.054
	(1.50)	(-0.10)	(1.48)	(16.12)	(-3.10)	(1.11)	(-1.23)	(-1.39)
LTR	-0.004	-0.048**	0.024	0.962***	-0.091	0.013	0.138	0.010
	(-0.22)	(-2.27)	(0.36)	(12.66)	(-0.85)	(1.48)	(0.60)	(0.39)
Finland								
LR	0.010**	0.011**	0.005	1.513***	-0.424***	-0.102	-0.043	-0.008
	(2.06)	(1.97)	(0.93)	(18.34)	(-2.90)	(-1.26)	(-1.26)	(-0.43)
STR	0.017	0.027**	0.008	1.352***	-0.355***	-0.034	-0.047	-0.017
	(1.63)	(2.18)	(0.69)	(16.16)	(-2.58)	(-0.43)	(-0.71)	(-0.41)
LTR	0.003	0.012*	-0.003	1.054***	0.054	-0.128	-0.031	0.003
	(0.50)	(1.84)	(-0.03)	(13.19)	(0.49)	(-1.65)	(-0.90)	(0.15)
Spain								
T-bill	-0.001	0.008	0.010	1.573***	-0.525***	0.001	0.051	-0.011
	(-0.14)	(0.75)	(0.98)	(17.11)	(-3.44)	(0.01)	(0.45)	(-0.47)
MMR	-0.004	0.021	0.003	1.117***	0.114	-0.247***	0.095	-0.023
	(-0.45)	(1.63)	(0.24)	(13.21)	(0.88)	(-2.96)	(0.73)	(-0.82)
LTR	0.002	0.012	0.011	1.251***	-0.219	-0.047	0.058	-0.019
	(0.22)	(0.79)	(0.81)	(14.35)	(-1.58)	(-0.54)	(0.37)	(-0.59)
Sweden								
BR	0.026	0.063*	0.013	0.926***	-0.016	0.039	0.363	-0.153
	(0.58)	(1.77)	(0.67)	(10.86)	(-0.14)	(0.49)	(0.78)	(-1.36)
STR	-0.026	0.014	-0.014	1.384***	-0.260*	-0.142*	0.403*	-0.049
	(-1.33)	(0.91)	(-1.64)	(17.28)	(-1.93)	(-1.80)	(1.98)	(-0.99)
LTR	0.000	0.011	0.001	1.369***	-0.325**	-0.060	-0.044	-0.002
	(0.01)	(0.50)	(0.04)	(16.27)	(-2.32)	(-0.71)	(-0.14)	(-0.02)
U.K.								
I.B	-0.081	0.025	0.007	0.402***	0.369***	0.114	1.864**	-0.763**
	(-0.93)	(0.57)	(0.57)	(4.97)	(4.69)	(1.48)	(2.25)	(2.01)
T-bill	-0.020	0.007	-0.021	1.488***	-0.597***	0.068	1.020*	-0.196*
	(-0.68)	(0.45)	(-0.92)	(18.90)	(-4.46)	(0.87)	(1.80)	(-1.73)
STR	-0.072*	-0.022	-0.002	1.425***	-0.648***	0.180**	0.932	-0.027
	(-1.93)	(-1.11)	(-0.07)	(18.19)	(-5.11)	(2.35)	(1.36)	(-0.17)
LTR	-0.072**	-0.006	0.006	1.203***	-0.336***	0.093	0.401	0.104
	(-2.13)	(-0.34)	(0.24)	(14.91)	(-2.75)	(1.20)	(0.70)	(0.75)

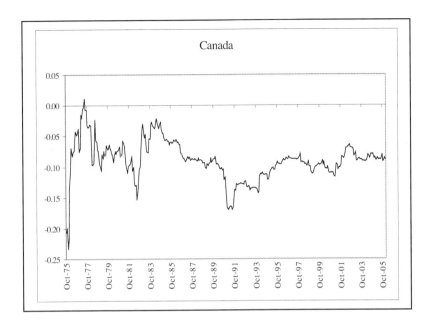

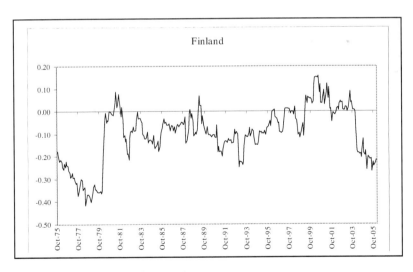

Figure 4.1 Time Variation of $\left(\phi_1 + \phi_2 \right)$.

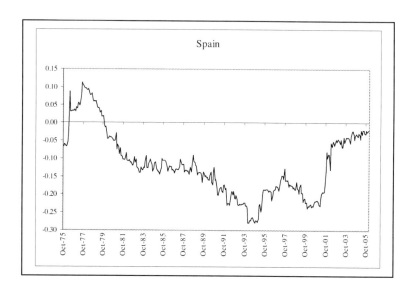

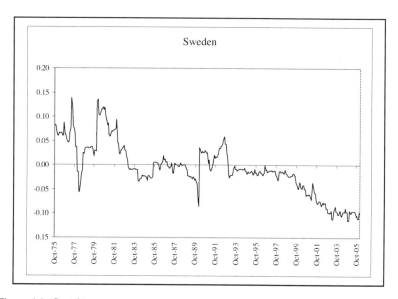

Figure 4.1 (Contd.)

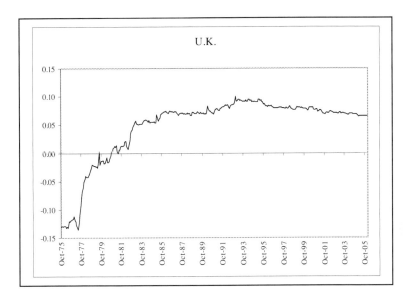

Figure 4.1 (Contd.)

Chapter 5

Interest Rate Model and Non-Linear Filtering

Heath-Jarrow-Morton (HJM, 1992) model of interest rates, based upon the specification of the forward interest rates, is the cornerstone of many aspects of modern fixed income market, both theory and applications. The main difficulty with this framework is that it requires user specification of a suitable volatility structure of the forward interest rate. Even in case of a simple structure, the model generates non-Markovian short term interest rate process. This makes it difficult for pricing applications for bonds, interest rate futures etc. This chapter will outline the strategies for transforming this non-Markovian structure to a Markovian structure by expansion of states. More importantly it will demonstrate how such a model may be estimated using stochastic filtering processes.

A class of volatility functions for the forward rate process is considered, which allows the bond price dynamics in the Heath-Jarrow-Morton (HJM) framework to be reduced to a finite-dimensional Markovian system. The use of this Markovian system in estimation of parameters of the volatility function via use of the Kalman filter is then discussed. Further, the Markovian system allows the link to be drawn between the HJM and the Vasicek and/or Cox-Ingersoll-Ross (CIR) frameworks for modeling the term structure of interest rates.

The content of this chapter relies on the article "Transformation of the Heath-Jarrow-Morton Models to Markovian System" published in The European Journal of Finance, volume 3 in 1997.

5.1 Background to HJM Model and the Related Literature

In the arbitrage-free approach to the modeling of the term structure of interest rates, the bond price dynamics are expressed in terms of an equivalent probability measure. As a consequence the drift and the diffusion coefficients of the stochastic differential equations describing the bond price dynamics become functions of a maturity-dependent volatility function. This approach has its origin in Ho and Lee (1986) but was most clearly articulated in Heath-Jarrow-Morton (HJM, 1992a). HJM (1992b) describe how their model can be used to price and hedge the entire interest rate derivative book of a financial institution thus offering a consistent approach in managing interest rate exposure.

The HJM approach requires the specification of the initial term structure in terms of forward rates and the volatility associated with these forward rates. The dynamics of the spot interest rate are then developed from those of the forward rate. The spot interest rate is also an important economic variable whose assessment determines the evolution of the bond prices. The major difficulty in implementation is that this spot interest rate is usually not path independent (i.e. it is non-Markovian) and the entire history of the term structure has to be carried, thus increasing the computational complexity.

The key unobserved input to this approach to term structure modeling is the aforementioned volatility of the forward rates. Many of the forms of the volatility functions reported in the literature have been chosen for analytical convenience rather than on the basis of empirical evidence. In fact apart from the study of HJM (1990), Flesaker (1993) and Amin and Morton (1994) there has not been a great deal of empirical research into the appropriate form of the volatility function to be used in the arbitrage free class of models. This is due to the fact that the non-Markovian nature of the stochastic dynamical system makes difficult application of standard econometric estimation procedures.

The non-Markovian feature has also made difficult the expression for prices of term-structure contingent claims in terms of partial differential equations. In the HJM approach these prices are expressed as expectation operators, under the equivalent martingale measure, of appropriate payoffs. Nowhere in the existing literature is it stated how to consistently turn this expectation operator into a partial differential equation. It is important to be able to do so in order to apply to the evaluation of interest rate sensitive contingent claims many of the recent computational advances as outlined in Wilmott et al. (1993). These techniques are the most appropriate to value various path-dependent options such as American, Asian, etc. but require an expression of the contingent claim price in terms of partial differential equation operators with appropriate boundary conditions.

This chapter assumes a form of the volatility function which is a product of the spot interest rate and a deterministic function of time having a specific but fairly general functional form. It is then shown how the spot rate process of the HJM framework leads to a finite-dimensional Markovian system. The dimension of the resultant system of stochastic differential equations is dependent on the exact form of the volatility function and it usually includes variables that are not readily observable. In order to establish an implementable estimation scheme, the Markovian system in state-space form requires application of nonlinear filtering techniques, e.g. the extended Kalman filter as discussed in Chapter 1 or Tanizaki (1993). This allows the prediction error decomposition form of the likelihood function to be constructed, which yields the parameter estimates when maximized.

The transformation to the Markovian form also allows easier comparison with other approaches, such as Vasicek (1977), Cox-Ingersoll-Ross (CIR) (1985) and Hull and White (1990). This is important in the sense that the existing literature seems unable to reconcile easily all of the alternative approaches to the modeling of the term structure of interest rates.

This chapter is laid out as follows. First, it reviews the essential framework of the HJM model relevant to this study. Second, it describes how the non-Markovian system for the instantaneous spot interest rate and bond price may be transformed to a Markovian system when the forward rate volatility is a deterministic function of time only. Third, it develops such a system when the volatility is a multiplicative function of the spot rate and the deterministic function of time. Fourth, it discusses how the Markovian system obtained in the previous sections can form the basis of a Kalman filter approach to the estimation of the parameters of the forward rate volatility function. Finally, it demonstrates how the Markovian representation allows the determination of a partial differential equation for the bond price which is the preference free version of the partial differential equation obtained by Vasicek (1977).

It should be pointed out that this development overlaps to some extent with that of Ritchken and Sankarasubramanian (1995) and Carverhill (1994). However, these authors do not focus on the Kalman filter and estimation issues nor on the link between the HJM and Vasicek/CIR approaches.

5.2 The Basic HJM Structure

We recall that the starting point of the HJM (1992a, b) model of the term structure of interest rates is the stochastic integral equation for the forward rate

$$f(t,T) = f(0,T) + \int_0^t \alpha(u,T)du + \int_0^t \sigma_f(u,T)dW(u), \ 0 \le u \le T \ (5.1)$$

where $f(t,T)$ is the forward rate at time t applicable to time T(> t). The noise term $dW(u)$ is the increment of a standard Wiener process generated by a probability measure Q. Note that in the interests of expositional simplicity we consider only one noise term impinging on the evolution of the forward rate. The functions $\alpha(u,T)$, $\sigma_f(u,T)$ are the instantaneous drift and volatility functions at time u of the forward rate $f(t,T)$. HJM show that the absence of riskless arbitrage opportunities implies that the drift term cannot be chosen arbitrarily but rather will be some function of the volatility function and the market price of interest rate risk. Furthermore, by an application of Girsanov's theorem, the explicit dependence on the market price of interest rate risk can be suppressed and the arbitrage-free stochastic integral equation for the forward rate can be written

$$f(t,T) = f(0,T) + \int_0^t \sigma_f(u,T) \int_u^t \sigma_f(u,s)dsdu + \int_0^t \sigma_f(u,T)d\tilde{W}(u) \ (5.2)$$

where $d\tilde{W}(u)$ is the increment of a standard Wiener process generated by an equivalent probability measure $\tilde{Q}$. It is possible to relate the probability measures Q and $\tilde{Q}$ by application of the Radon-Nikodym theorem but the details of this are not necessary for the purposes of our discussion.

It is then a simple matter to deduce that the instantaneous spot rate of interest $(r(t) = f(t,t))$ satisfies the stochastic integral equation

$$r(t) = f(0,t) + \int_0^t \sigma_f(u,t) \int_u^t \sigma_f(u,y)dydu + \int_0^t \sigma_f(u,t)d\tilde{W}(u). \quad (5.3)$$

Alternatively equation (5.3) can be expressed as the stochastic differential equation,

$$dr(t) = \left[f_2(0,t) + \frac{\partial}{\partial t}\int_0^t \sigma_f(u,t)\int_u^t \sigma_f(u,y)dydu \right]dt$$
$$+ \left[\int_0^t \frac{\partial \sigma_f(u,t)}{\partial t}d\tilde{W}(u) \right]dt + \sigma_f(t,t)d\tilde{W}(t). \quad (5.4)$$

Application of Ito's lemma shows that the price of a pure discount bond paying \$1 at time T, $P(t,T)$, evolves according to:

$$dP(t,T) = r(t)P(t,T)dt - P(t,T)\int_t^T \sigma_f(t,u)dud\tilde{W}(t)$$
$$= r(t)P(t,T)dt + \sigma_B(t,T)P(t,T)d\tilde{W}(t). \quad (5.5)$$

It is at times convenient to deal with the log of the bond price $B(t,T) = \log P(t,T)$. This quantity, by Ito's lemma, satisfies,

$$dB(t,T) = \left[r(t) - \sigma_B^2(t,T) \right]dt + \sigma_B(t,T)d\tilde{W}(t). \quad (5.6)$$

If we define the accumulated money market account,

$$A(t) = \exp\left[\int_0^t r(y)dy \right] \quad (5.7)$$

then the relative bond price,

$$Z(t,T) = \frac{P(t,T)}{A(t)}, \quad 0 \le t \le T$$

satisfies the stochastic differential equation,

$$dZ(t,T) = \sigma_B(t,T)Z(t,T)d\tilde{W}(t). \quad (5.8)$$

Equation (5.8) implies that under the probability measure $\tilde{Q}$, $Z(t,T)$ is a martingale. It is then a simple matter to derive the well known result,

$$P(t,T) = \tilde{E}_t\left[\exp\left\{-\int_0^t r(y)dy\right\}\right] \qquad (5.9)$$

where $\tilde{E}_t$ denotes the mathematical expectation operator with respect to the probability measure $\tilde{Q}$. Equation (5.9) is the desired preference-free expression derived by HJM (1992a, b) for the term structure of interest rates. A more detailed summary of the key steps in the derivation of the HJM results is given in Appendix 5.1.

The principal difficulty in implementing and estimating HJM models arises from the non-Markovian noise term in the stochastic integral equation (5.3) for r(f). This manifests itself in the third component of the drift term of the stochastic differential equation (5.4). This component depends on the history of the noise process from time 0 to current time t. Our aim now is to investigate particular functional forms of $\sigma_f(t,T)$ that allow the non-Markovian representation of r(t) and P(t, T) to be reduced to a finite-dimensional Markovian system of stochastic differential equations. We investigate volatility functions of the forward rate which are:
Deterministic functions of time of the form

$$\sigma_f(t,T) = p_n(T-t)e^{-\lambda(T-t)} \equiv \sigma(t,T) \qquad (5.10)$$

where $p_n(u)$ is the polynomial of the form $p_n(u) = a_0 + a_1u + \dots + a_nu^n$, and λ, a_i's are parameters to be estimated.
Product of a deterministic function of time and the instantaneous spot rate of interest, i.e.

$$\sigma_f(t,T) = \sigma(t,T)G(r(t)) \qquad (5.11)$$

where G is an appropriately well-behaved function.

The motivation for the class of volatility functions chosen is to allow a high degree of flexibility in modeling the wide range of shapes of the yield curve by virtue of the polynomial in the deterministic part. The functional dependence on r(t) in equation (5.11) allows for dependence on driving stochastic variables. Ideally we would like also to allow for a functional dependence on f(t, T), as in Amin and Morton (1994). However, this requires an extension of the approach we advocate here because in this case the state space representation that we derive below becomes infinite-dimensional. Nevertheless, the approach adopted here provides the framework in which these further developments can be embedded. Essentially this extension requires application of infinite-dimensional filtering techniques as discussed in Falb (1967).

5.3 Forward Rate Volatility: Deterministic Function of Time

We consider in this section the case where the forward rate volatility is a deterministic function of time. Our principal result in this section can be stated as follows:

Proposition 1:
   If the forward rate volatility function assumes the form,

$$\sigma_f(v,t) = p_n(t-v)e^{-\lambda(t-v)} \qquad (5.12)$$

then the instantaneous spot interest rate r(f) and log bond price B(t, T) are determined by the (n+2)-dimensional Markovian stochastic differential equation system,

$$dS(t) = \left[J(t) + HS(t) \right] dt + V(t) d\tilde{W}(t) \qquad (5.13)$$

where,

$$S(t) = \left[B(t,T), r(t), Z_0(t), ..., Z_{n-1}(t) \right]^T \qquad (5.14)$$

and $Z_i(t)$, i=0,1,2,...,n-1 are subsidiary stochastic variables describing the statistical characteristics of the history of the noise process up to time t. The (time-dependent) vectors J(t), V(t) and (constant) matrix H depend nonlinearly on the parameters $a_0, a_1, ..., a_n, \lambda$ and are defined in Appendix 5.2.

Proof of Proposition 1:
   The proof is given in the Appendix 5.2. It relies on defining quantities,

$$Z_i(t) = \int_0^t (t-v)^i e^{-\lambda(t-v)} d\tilde{W}(v)$$

which are measures of the statistical properties of the noise process. The fact that these quantities satisfy the finite-dimensional stochastic dynamical system,

$$dZ_0(t) = -\lambda Z_0(t) + d\tilde{W}(t)$$

$$dZ_i(t) = \left[iZ_{i-1}(t) - \lambda Z_i(t) \right] dt, \; i=1,2,...,n$$

is the crucial observation which allows us to reduce the original non-Markovian system to Markovian form of the dimension indicated.

For the purposes of estimation of HJM models we need to consider the stochastic differential equation for the bond price P(t,T). However, we have considered instead the stochastic differential equation for the log of the bond price since the Markovian system developed in equation (5.13) turns out to be linear in the state variables. This latter result will be most convenient from the point of view of implementing estimation procedures (Bhar and Chiarella, 1995b).

Since the volatility vector V(t) is independent of the state variables and is a function of time only, the stochastic dynamic system (5.13) is Gaussian. The quantities $Z_i(t)$ (for all i) which summarize the history of the noise process, as well as the instantaneous spot rate of interest r(t) are not readily observable. Certainly there exists a one-to-one mapping between the $Z_i(t)$ and yields drawn from the term structure and it would be possible to take (n+1) such points to tie down the $Z_i(t)$. While this would in principle provide an alternative estimation procedure it has the practical drawback that it would require continuous observations of the term structure. Many empirical studies attempt to proxy r(t) by some short-term rate, e.g. 30-day treasury bill rates. Thus, in developing estimation techniques we need also to consider the observation vector (which in this case reduces to a scalar)

$$Y(t) = CS(t), \; C=[1,0,.....,0]. \tag{5.15}$$

The system (5.13) with the observation vector (5.14) is now in a form to which we are able to apply Kalman filter estimation techniques to form the log-likelihood function and hence estimate the parameters specifying the volatility function.

For later purpose it is of interest to focus on the special case $a_0 \neq 0$ and $a_1 = a_2 = ... = a_n = 0$, so that,

$$\sigma_f(v,t) = a_0 e^{-\lambda(t-v)}. \tag{5.16}$$

This form of the forward rate volatility corresponds to that of the Vasicek (1977) model. In this case there is only one subsidiary stochastic variable, namely $Z_0(t)$. However, it does not directly influence the drift term of the stochastic differential equation for r(t), which then becomes,

$$dr(t) = \left[D^*(t) - \lambda r(t) \right] + a_0 d\tilde{W}(t) \tag{5.17}$$

where,

$$D^*(t) = D(t) + \lambda f(0,t) + \lambda \int_0^t \sigma(v,t) \int_v^t \sigma(v,y) dy dv.$$

The stochastic differential equation (5.17) is of the same mean-reverting form as that employed by Vasicek. The difference being that the short-run mean D*(t), because of the arbitrage-free derivation, is dependent upon the initial forward rate curve and the function describing the volatility of the forward rate.

5.4 Forward Rate Volatility: Stochastic

In this section we shall assume that the forward rate volatility has the following form:

$$\sigma_f(t,T) = \sigma(t,T) G\big(f(t,t)\big) = \sigma(t,T) G\big(r(t)\big) \tag{5.18}$$

where G is a well-behaved function and a typical example would be,

$$G\big(r(v)\big) = r(v)^\gamma, \gamma \geq 0. \tag{5.19}$$

The reasonableness of the functional specification in equation (5.18) remains an empirical issue. Our main purpose in introducing it here is to provide yet another example of a specification under which the HJM approach yields a finite-dimensional Markov process for the state variables. A secondary purpose is to help clarify the link between the HJM approach and that of Vasicek/CIR.

The argument leading to the arbitrage-free expression for the stochastic integral equation driving the forward rate remains the same, so that we have,

$$f(t,T) = f(0,T) + \int_0^t \sigma_f(\upsilon,T) \int_\upsilon^T \sigma_f(\upsilon,y) dy d\upsilon + \int_0^t \sigma_f(\upsilon,T) d\tilde{W}(\upsilon)$$

$$= f(0,T) + \int_0^t G(r(\upsilon))^2 \sigma(\upsilon,T) \int_\upsilon^T \sigma(\upsilon,y) dy d\upsilon \qquad (5.20)$$

$$+ \int_0^t G(r(\upsilon))\sigma(\upsilon,T) d\tilde{W}(\upsilon).$$

Equation (5.20) may also be written in a stochastic differential form,

$$df(t,T) = \left[G(r(t))^2 \sigma(t,T) \int_t^T \sigma(t,y) dy \right] dt$$

$$+ G(r(t))\sigma(t,T) d\tilde{W}(t). \qquad (5.21)$$

The corresponding stochastic differential equation for the instantaneous spot rate is given by,

$$r(t) = f(0,t) + \int_0^t G(r(\upsilon))^2 \sigma(\upsilon,t) \int_\upsilon^t \sigma(\upsilon,y) dy d\upsilon$$

$$+ \int_0^t G(r(\upsilon))\sigma(\upsilon,t) d\tilde{W}(\upsilon) \qquad (5.22)$$

which has also the differential representation as,

$$dr(t) = \left[f_2(0,t) + \frac{\partial}{\partial t} \int_0^t G(r(\upsilon))^2 \sigma(\upsilon,t) \int_\upsilon^t \sigma(\upsilon,y) dy d\upsilon \right] dt$$

$$+ \left[\int_0^t G(r(\upsilon))\sigma_2(\upsilon,t) d\tilde{W}(\upsilon) \right] dt + \qquad (5.23)$$

$$+ G(r(t))\sigma(t,t) d\tilde{W}(t).$$

Using equations (5.20) and (5.22) we obtain,

$$\frac{dP(t,T)}{P(t,T)} = r(t)dt + \left[-\int_t^T G(r(t))\sigma(t,\upsilon) d\upsilon \right] d\tilde{W}(t)$$

$$= r(t)dt + G(r(t))\sigma_B(t,T) d\tilde{W}(t). \qquad (5.24)$$

Again, it follows from Ito's lemma that $B(t,T) = \ln P(t,T)$,

$$dB(t,T) = \left[r(t) - 0.5G\left(r(t)\right)^2 \sigma_B^2(t,T) \right] dt$$
$$+ G\left(r(t)\right)\sigma_B(t,T) d\tilde{W}(t).$$

(5.25)

We note again that the third component of the drift term, equation (5.23) for r(t) is non-Markovian in general. If this stochastic differential equation for r(t) can be reduced to Markovian form then we would have a Markovian system determining the state variables r(f), P(t,T) (or B(t,T)).

It is much easier to see the overall structure of the approach to reducing (5.22) to Markovian form by focusing on the following special case.

Proposition 2:
    When the forward rate volatility function is of the form,

$$\sigma_f(t,T) = \sigma(t,T)G\left(r(t)\right) = \left[a_0 + a_1(T-t) \right] e^{-\lambda(T-t)} G\left(r(t)\right) \quad (5.26)$$

the instantaneous spot rate r(t) and log bond price B(t,T) are determined by the finite dimensional Markovian system,

$$dS(t) = \left[J(r,t) + HS(t) \right] dt + V(r,t) d\tilde{W}(t) \qquad (5.27)$$

where, $S(t) = \left[B(t,T), r(t), Z_0(t), \phi_{0,2}(t) \right]^T$ for n=0. For n=1, we get, $S(t) = \left[B(t,T), r(t), Z_0(t), \phi_{0,1}(t), \phi_{0,2}(t), \phi_{1,2}(t), \phi_{2,2}(t) \right]^T$. Here, $Z_0(t)$, $\phi_{i,j}(t), i=0, j, j=1,2$ are subsidiary stochastic variables describing statistical characteristics of the history of the noise process up to time t. The time and state-dependent vectors J(r,t), V(r,t) and the constant matrix H are defined in Appendix 5.3

Proof of Proposition 2:
    The proof is given in Appendix 3 and relies upon the quantities,

$$Z_i(t) = \int_0^t G\left(r(u)\right)(t-u)^i e^{-\lambda(t-u)} d\tilde{W}(u), \quad i=0,1$$

which in turn satisfy the finite dimensional stochastic dynamic system,

$$dZ_0(t) = -\lambda Z_0(t) dt + G\left(r(t)\right) d\tilde{W}(t)$$

$$dZ_1(t) = \left[Z_0(t) - \lambda Z_1(t) \right] dt$$

and,

$$\phi_{k,j} = \int_0^t G\big(r(u)\big)^2 (t-u)^k e^{-j\lambda(t-u)} du, \ k=0,j \text{ and } j=1,2 \ .$$

The differential forms of the above integral equation are given in Appendix 5.3. These quantities measure statistical properties of the time path of interest rates. As with Proposition 1 the fact that they satisfy a finite-dimensional stochastic system is the crucial observation which allows us to reduce the original non-Markovian system to Markovian form of appropriate dimension.

The observation vector in this case is similar to equations (5.15) with appropriate dimension of C.

Let us now consider some specific examples. Consider first of all the case when the deterministic function of time $\sigma(t,T)$ is a constant and G is the power function specified in equation (5.19), i.e. $a_1 = 0$, $\lambda=0$, so that,

$$\sigma_f(\upsilon,T) = a_0 r(\upsilon)^\gamma \ . \tag{5.28}$$

This form of the volatility for the forward rate allows us to see the link with the Vasicek/CIR approach. It follows that the Markovian stochastic differential equation system then becomes

$$dr(t) = \Big[f_2(0,t) + a_0^2 \phi_{0,2}(t) \Big] dt + a_0 r^\gamma(t) d\tilde{W}(t)$$
$$d\phi_{0,2}(t) = r^{2\gamma}(t) dt \ . \tag{5.29}$$

The volatility function for the instantaneous spot rate of interest in equation (5.29) is of the same form as in the Vasicek/CIR class of models.

Another example is obtained by setting $a_1 = 0$, $\lambda \neq 0$ so that,

$$\sigma_f(t,T) = a_0 G\big(r(t)\big) e^{-\lambda(T-t)} \ . \tag{5.30}$$

In this case the system corresponding to (5.29) becomes,

$$dr(t) = \Big[f_2(0,t) + \lambda f(0,t) - \lambda r(t) + a_0^2 \phi_{0,2}(t) \Big] dt$$
$$a_0 G\big(r(t)\big) d\tilde{W}(t) \ , \tag{5.31}$$

$$d\phi_{0,2}(t) = \Big[G\big(r(t)\big)^2 - 2\lambda \phi_{0,2}(t) \Big] dt \ . \tag{5.32}$$

This is the two-dimensional, Markovian system obtained by Ritchken and Sankarasubramanyan (1995).

Proposition 2 gives a representation for the special cases $n = 0, 1$. For the case of general $n \geq 1$ we can state:

Proposition 3:

When the forward rte volatility has the following functional form,

$$\sigma_f(t,T) = p_n(T-t)e^{-\lambda(T-t)}G(r(t))$$

the instantaneous spot rate $r(t)$ and bond price are determined by a Markovian system of dimension $0.5 \times (n^2 + 7n + 8)$ and has the form:

$$dS(t) = [J(r,t) + HS(t)]dt + V(r,t)d\tilde{W}(t),$$

$$S(t) = [B(t,T), r(t), Z_0(t), ..., Z_{n-1}(t), \phi_{0,1}(t), ..., \phi_{n+1,n+1}(t)]^T$$

and $Z_{i,t}, i = 0,1,...,n-1$, $\phi_{j,k}(t)$, $0 < k \leq n+1$, $0 \leq j \leq k$ are subsidiary stochastic variables describing statistical characteristics of the history of the noise process up to time t. The time and state dependent vectors $J(r,t)$, $V(r,t)$ and the constant matrix H can be set up from the procedure described in Appendix 5.4.

Proof of Proposition 4:

This proof is given in Appendix 5.4.

The dimension of the system increases rapidly with n. For n=1, the dimension is 8, and for n=2 it is 13. It is thus clear that the computational burden, even for n=2 is extremely high. But for special cases such as equations (5.28) and (5.30) for forward rate volatility, a three-dimensional system is involved which is fairly simple to implement.

It is important to point out that the finite-dimensional state-space representation in Propositions 1, 2, 3 are not unique. For certain values of n it is possible to find relationships between the subsidiary variables $Z_i(t)$, $\phi_{i,j}(t)$ that allow a reduction in the dimensionality of the system. For instance in the case n=0, Proposition 3 indicates that we are dealing with a four-dimensional system in $B(t,T), r(t), Z_0(t), \phi_{0,2}(t)$.

5.5 Estimation via Kalman Filtering

In order to estimate the parameters of the volatility function of the forward rate based upon a series of observations on bond prices (or log bond prices), the spot instantaneous interest rate enters the process. This is not a readily observable variable in the market. Besides, the path-dependent nature of the instantaneous spot rate in the HJM framework increases the burden of computation. Hence, the need for transformation to a Markovian form arises. The previous two sections describe how this can be achieved for a fairly general class of volatility function. In this section we outline how the Markovian state-space representation obtained in the previous two sections allows application of Kalman filtering methodology to the problem of estimating the parameters used to specify the forward rate volatility function. Further details of this approach together with implementation to market data can be found in Bhar and Chiarella (1995a).

It is important to stress that the state space representation is merely a re-expression of the expectation operator form (5.9). Furthermore, both representations are under the equivalent measure $\tilde{Q}$ and are therefore preference free. The expectation operator form (5.9) has been the starting point of those empirical studies based on preference-free models such as Flesaker (1993) and Amin and Morton (1994). The intuitive reason why it is appropriate to do empirical analysis under this measure is that the future price evolution is with respect to the currently observed initial term structure. This is in contrast to the equilibrium class of models, such as Cox et al. (1985), which seek to explain both the current term structure as well as the future price evolution as a function of a (small) number of dynamic economic state variables. This class of models naturally involves the market price of interest rate risk and hence their representations are under the original measure Q and necessarily involve estimating the market price of interest rate risk. Hence our state space representation under the equivalent measure $\tilde{Q}$ is the appropriate form to use when the dynamic evolution is with respect to the current initial term structure.

The simultaneous stochastic differential equations given by (5.13) and (5.27) have the general structure

$$dS(t) = F(S(t); \Theta)dt + V(S(t), t; \Theta)d\tilde{W}(t) \qquad (5.33)$$

with the observation vector,

$$Y(t) = CS(t). \qquad (5.34)$$

In the general case, F and V may be non-linear in the state variables and Θ represents the parameters of the volatility function to be estimated.

Expanding the drift term in equation (5.33) in a first-order Taylor series about each observation $S(t_k)$, t_k ($k = 1,...,N$), concentrating on the interval $\delta_k = t_k - t_{k-1}$ and approximating the volatility vector over this subinterval by its value at the beginning of the interval, the system (5.33) reduces to the linear system (1 is the size of the state vector S):

$$dS(t) = \left[a_k + A_k S(t) \right] dt + V_k d\tilde{W}(t) \qquad (5.35)$$

where,

$$A_k = \left\{ \frac{\partial F_i}{\partial S_j} \bigg|_{S_k} \right\}, \; i, j = 1, 2, ..., l \text{ and}$$

$$a_k = F_k - A_k S_k, \; F_k \equiv F(S(t_k); \Theta), \; S_k \equiv S(t_k), \; V_k = V(S_k, t_k).$$

The crucial point is that this procedure leaves V_k constant within the interval $\delta_k = t_k - t_{k-1}$. In this interval, since equation (5.35) is linear in $S(t)$, the solution of equation (5.35) is given by,

$$S_{k+1} = \exp\left[A_k \delta_k \right] S_k + \int_0^{\delta_k} \exp\left[A_k (\delta_k - \upsilon) a_k d\upsilon \right]$$
$$+ \int_0^{\delta_k} \exp\left[A_k (\delta_k - \upsilon) V_k d\tilde{W}(\upsilon) \right]. \qquad (5.36)$$

Since $\tilde{W}$ is Wiener process, S_{k+1} has a conditional normal distribution within $\delta_k = t_k - t_{k-1}$ and the first two moments are:

$$E[S_{k+1}] = \exp\left[A_k \delta_k \right] S_k + \int_0^{\delta_k} \exp\left[A_k (\delta_k - \upsilon) a_k d\upsilon \right],$$
$$\text{cov}[S_{k+1}] = \int_0^{\delta_k} \exp\left[A_k \upsilon \right] V_k V_k^T \exp\left[A_k^T \upsilon \right] d\upsilon. \qquad (5.37)$$

Under the assumption of a normal distribution as incorporated in (5.37) the transition probability density function for the state vector S_k to S_{k+1} can be written for a given set of observations with the help of the Kalman filter algorithm described in Chapter 1.

Following the steps outlined in Chapter 1 the likelihood function can be developed as a function of the parameter vector Θ. The parameter vector can be estimated by maximizing this likelihood function with a suitable numerical optimization technique. Further details and applications to empirical data are discussed in Bhar and Chiarella (1995a) as well as in Tanizaki (1993).

The issue of identifiability is important in the estimation of the model parameters, since lack of identification may lead to practical difficulties. This concept centers on the notion of observational equivalence between two or more structures of the model (Wall, 1987). For a given form of the volatility function, the state space representation used in this section ensures a unique correspondence between the parameters of forward rate volatility function and the error covariance. Thus, the forecast error decomposition form of the likelihood function is well defined in terms of the unknown parameters.

5.6 Preference-Free Approach to Bond Pricing

Both the HJM and Vasicek approaches to the modeling of the term structure are based on arbitrage arguments. The HJM approach has the important advantage that it is preference free and contains all of the information impounded in the current forward rate curve. Another important distinction between the two approaches is the mathematical expression for bond price. In HJM, because of their use of martingale ideas, the bond price is expressed cis an expectation, whose transition probability density function is driven by a non-Markovian stochastic dynamical system. In Vasicek, the bond price is determined by a partial differential equation, which Vasicek shows can also be expressed as an expectation although both forms involve the market price of interest rate risk. Furthermore the Vasicek approach does not make use of the information in the current forward rate curve, although Vasicek does show that given the particular functional form that he assumes for the market price of risk function, the market price of risk parameter can be related to the slope of the current forward rate curve at maturity.

The relationship between these two arbitrage based approaches to term structure modeling still remains unclear since one does not seem to find easily in the literature a partial differential equation representation of the bond price in the HJM framework. The Markovian representation of the spot rate derived here allows the derivation of a preference-free differential equation for the bond price which bears a strong resemblance to the one obtained from the Vasicek approach.

We have seen in Section 5.3 that assuming the Vasicek form for the forward rate volatility, the spot rate of interest in an arbitrage free economy is driven by the stochastic differential equation (5.17), which we rewrite here,

$$dr(t) = \left[D^*(t) - \lambda r(t)\right] + a_0 d\tilde{W}(t).$$ (5.38)

The stochastic differential equation (5.38) has associated with it the Kolmogorov backward equation for the transition probability density $\pi(r(T), T \mid r(t), t)$ (i.e. the probability of observing r(T) at time T conditional on r(f) at time t),

$$\frac{1}{2} a_0^2 \frac{\partial^2 \pi}{\partial r^2} + \left[D^*(t) - \lambda r(t)\right] \frac{\partial \pi}{\partial r} + \frac{\partial \pi}{\partial t} = 0$$ (5.39)

where t ranges, say, between the limits $T_c < t < T$. The initial time T_c could for example be the maturity date of an option on a bond maturing at T.

It is more convenient to introduce the elliptic partial differential operator K and write (5.39) more succinctly as,

$$K\pi + \frac{\partial \pi}{\partial t} = 0, \ T_c < t < T.$$ (5.40)

We also recall that (5.39) is solved subject to the initial condition,

$$\pi(r(T), T \mid r, T) = \delta(r(T) - r)$$ (5.41)

where δ is the Dirac delta function.

Given the transition probability density function n we can in principle calculate the expectation in (5.9). However this calculation is not so simple as we need to calculate the expectation not of a function of the state variable r(f), but rather of a functional of that variable, namely,

$$\exp\left(-\int_t^T r(s)ds\right).$$

The technique that allows us to calculate the expectations of such functional is the Feynman-Kac theorem which is discussed in Gikhman and Skorokhod (1965). Application of this result allows us to state that the expectation of the functional in (5.9) satisfies the partial differential equation,

$$KP + \frac{\partial P}{\partial t} - rP = 0, \ T_c \leq t \leq T \tag{5.42}$$

subject to the initial condition, $P(T,T) = 1$.

Equation (5.42) should be compared with the partial differential equation for the bond price which is derived by Vasicek. They differ only in the coefficient of the $\partial P/\partial r$ term, which in (5.42) is preference-free and dependent on the volatility function, whereas in Vasicek this coefficient involves the arbitrarily specified drift term in the process for r, the preference-dependent market price of risk parameter and a volatility parameter.

Equation (5.42) should be viewed as the preference-free version of the Vasicek partial differential equation. It is also possible by following the approach we have adopted to obtain a corresponding preference-free partial differential equation when one adopts the Cox-Ingersoll-Ross (1985) square root process for the volatility of the spot rate. Furthermore it is possible to draw a link to the approach of Hull and White (1990) who assume a time varying drift term in the process for r(f) and by various manipulations express it in terms of the current forward rate curve. These manipulations are essentially an indirect way of arriving at the expression $D^*(t)$ in equation (5.17). Many of these points are discussed in detail and elaborated upon in Chiarella and El-Hassan (1996). However, our main purpose here has been to show how the Markovian representation of HJM allows preference-free forms of the Vasicek (and indeed Cox-Ingersoll-Ross) partial differential equations to be obtained.

It has been shown by Ritchken and Sankartisubramanian (1995) that in the case of volatility functions of the form (5.10) it is possible to obtain a closed form solution for the bond price in equation (5.42). This fact has been employed by Chiarella and El-Hassan (1996) in using the above framework to calculate numerically values of American bond options.

5.7 Concluding Remarks

The Heath-Jarrow-Morton (1992a, b) model of the term structure of interest rates has the desirable property of being preference-free. It also has the characteristic that it allows the stochastic dynamics of the instantaneous spot rate of interest and bond prices to be non-Markovian, a characteristic which makes it more general (at least in its dynamic evolution structure) than models which have preceded it. It is also this non-Markovian characteristic which is the principal difficulty in implementing and estimating Heath-Jarrow-Morton models.

In this paper we have shown that specification of the forward rate volatility as a certain deterministic function of time multiplied by a function of the instantaneous spot rate of interest allows the dynamics of the Heath-Jarrow-Morton model of the term structure of interest rates to be expressed as a finite-dimensional Markovian stochastic dynamic system. This dynamical system may be discretized to a form whose estimation is conveniently approached using the full array of well-developed Kalman filter techniques. The approximation procedure leading to the discretization in equation (5.36) is about the crudest that can be employed. Better discretization schemes are possible and these are discussed in Bhar and Chiarella (1995b). Our main aim in this paper has been to show how it is possible to go from the HJM formulation of the term structure to an estimable discretized form under a reasonable specification of the volatility of the forward rate process.

A further use of the Markovian representation we have developed is to allow representation of the partial differential operator of the Kolmogorov equation for the transition probability density under the equivalent martingale measure.

Having this partial differential operator allows us to write down the partial differential equations satisfied by the bond price and other contingent claims (e.g. bond options). Such a representation provides an alternative starting point for the numerical implementation of the Heath-Jarrow-Morton framework. Some of these issues are explored in Chiarella and El-Hassan (1996), particularly in relation to the evaluation of American bond options within the HJM framework.

A topic for further investigation is to develop techniques which allow the forward rate volatility function to be a function of the forward rate itself. In this case it does not seem possible to obtain a reduction to a finite dimensional Markovian system and it becomes necessary to view the estimation problem as one of Kalman filtering in infinite-dimensional space.

Appendix 5.1 Arbitrage-Free SDE for the Bond Price

This appendix merely summarizes and slightly reorders the key results of HJM (1992a) in the special case of one factor uncertainty. Suppose, the forward rate is given by the stochastic integral equation,

$$f(t,T) = f(0,T) + \int_0^t \alpha(v,T)dv + \int_0^t \sigma_f(v,T)dW(v), \ 0 \le t \le T \quad (5.43)$$

or alternatively, by the stochastic differential equation,

$$df(t,T) = \alpha(t,T)dt + \sigma_f(t,T)dW(t). \quad (5.44)$$

It follows easily that the instantaneous spot rate $r(t) \equiv f(t,t)$ is given by,

$$r(t) = f(0,T) + \int_0^t \alpha(v,T)dv + \int_0^t \sigma_f(v,T)dW(v). \quad (5.45)$$

The bond price at time t is related to the forward rate by,

$$P(t,T) = \exp\left(-\int_t^T f(t,s)ds\right), \ 0 \le t \le T. \quad (5.46)$$

By the use of Fubini's theorem for stochastic integrals and application of Ito's lemma, HJM show that the bond price must satisfy the stochastic differential equation,

$$dP(t,T) = [r(t) + b(t,T)]P(t,T)dt + a(t,T)P(t,T)dW(t) \quad (5.47)$$

and,

$$a(t,T) = -\int_t^T \sigma_f(t,v)dv, \ b(t,T) = -\int_t^T \alpha(t,v)dv + \frac{1}{2}a(t,T)^2.$$

A quantity of interest is the accumulated money account,

$$A(t) = \exp\left[\int_0^t r(y)dy\right] \quad (5.48)$$

which is then used to define the relative bond price,

$$Z(t,T) = \frac{P(t,T)}{A(t,T)}, \ 0 \le t \le T. \quad (5.49)$$

The relative bond price satisfies the stochastic differential equation,

$$dZ(t,T) = b(t,T)Z(t,T)dt + a(t,T)Z(t,T)dW(t). \quad (5.50)$$

We know from standard arbitrage arguments (Vasicek, 1977) that in order that riskless arbitrage opportunities between bonds of differing maturities do not exist, then the instantaneous bond return in (5.47) must satisfy,

$$\frac{\left[r(t)+b(t,T)P(t,T)\right]-r(t)P(t,T)}{a(t,T)P(t,T)}=-\phi(t) \qquad (5.51)$$

where $\phi(t)$ is the market price of interest rate risk and (5.51) simplifies to,

$$b(t,T)=\phi(t)a(t,T)=0 . \qquad (5.52)$$

Using the definitions in (5.47), this last equation may be written explicitly as,

$$\int_t^T \alpha(t,v)dv -\frac{1}{2}\left(\int_t^T \sigma_f(t,v)dv\right)^2 +\phi(t)\sigma_f(t,T)=0 \qquad (5.53)$$

i.e.

$$\alpha(t,T)=-\sigma_f(t,T)\left[\phi(t)-\int_t^T \sigma_f(t,v)dv\right], \qquad (5.54)$$

which is HJM equation (18).

Up to this point HJM have not done anything different from the standard arbitrage approach of Vasicek. They do, however, interpret equation (5.52) in a different way. In the standard arbitrage approach of Vasicek, (5.53) becomes a partial differential equation for the bond price as a function of the assumed driving state variable (usually the instantaneous spot rate). In the HJM approach, this condition (5.53) becomes a relationship which ties down the drift of the forward rate process as a function of the volatility and the market price of interest rate risk, which is what equation (5.54) asserts. Note that by use of (5.53) the stochastic differential equations for P(t,T) and Z(t,T) become,

$$dP(t,T)=\left[r(t)-\phi(t)a(t,T)\right]P(t,T)dt +a(t,T)P(t,T)dW(t) \qquad (5.55)$$

and,

$$dZ(t,T)=-\phi(t)a(t,T)Z(t,T)dt +a(t,T)Z(t,T)dW(t). \qquad (5.56)$$

At the same time, by integrating (5.54) over (0, t) the stochastic integral equation for r(t) becomes,

$$r(t) = f(0,t) + \int_0^t \sigma_f(v,t) \int_v^t \sigma_f(v,y) dv dy$$
$$- \int_0^t \sigma_f(v,t)\phi(v) dv + \int_0^t \sigma_f(v,t) dW(v). \tag{5.57}$$

Note that in (5.57) we use the relation,

$$\int_0^t \alpha(v,t) dv = -\int_0^t \sigma_f(v,t)\phi(v) dv + \int_0^t \sigma_f(v,t) \int_v^t \sigma_f(v,y) dy dv.$$

The key advance in the HJM approach is the observation that (5.55) – (5.57), by use of Girsanov's theorem, can be written in terms of a different Brownian motion generated by an equivalent martingale measure. Thus, if we define new Brownian motion $\tilde{W}(t)$ by,

$$\tilde{W}(t) = W(t) - \int_0^t \phi(s) ds, \text{ i.e. } d\tilde{W}(t) = dW(t) - \phi(t) dt \tag{5.58}$$

then equations (5.55) – (5.57) become,

$$dP(t,T) = r(t)P(t,T) dt + a(t,T)P(t,T) d\tilde{W}(t), \tag{5.59}$$

$$dZ(t,T) = a(t,T)Z(t,T) d\tilde{W}(t)\overset{\bullet}{,} \tag{5.60}$$

$$r(t) = f(0,t) + \int_0^t \sigma_f(v,t) \int_v^t \sigma_f(v,y) dy dv + \int_0^t \sigma_f(v,t) d\tilde{W}(v). \tag{5.61}$$

The essential characteristic of the reformulated stochastic differential and integral equations (5.59) – (5.61) expressed in terms of Brownian motion under the equivalent probability measure is that the empirically awkward market price of risk term $\phi(t)$ is eliminated from explicit consideration.

If we write $\tilde{E}_t$ to denote mathematical expectation at time t with respect to the equivalent probability measure, then,

$$\tilde{E}_t[dZ(t,T)] = 0. \tag{5.62}$$

This last equation implies that $Z(t,T)$ is a martingale, i.e.

$$Z(t,T) = \tilde{E}_t[Z(T,T)] \tag{5.63}$$

or in term of bond price,

$$P(t,T) = \tilde{E}_t \left[\frac{A(t)}{A(T)} \right] = \tilde{E}_t \left(\exp \left\{ -\int_t^T r(y) dy \right\} \right). \tag{5.64}$$

The expectation in equation (5.64) could be calculated by numerically simulating (5.61), although close form analytical expressions may be obtained with appropriate assumptions on $\sigma_f(t,T)$.

Appendix 5.2: Proof of Proposition 1

The proof relies on a reduction of stochastic integrals involving exponential and polynomial functions by successive use of Fubini's theorem.

The assumed functional form is,

$$\sigma_f(v,t) = \left[a_0 + a_1(t-v) + ... + a_n(t-v)^n \right] e^{-\lambda(t-v)} \tag{5.65}$$

and define the subsidiary stochastic variables,

$$Z_i(t) \equiv \int_0^t (t-v)^i e^{-\lambda(t-v)} d\tilde{W}(v), \; i = 0,1,2,...,n \;, \tag{5.66}$$

then,

$$\int_0^t \sigma_f(v,t) d\tilde{W}(v) = a_0 Z_0(t) + a_1 Z_1(t) + ... + a_n Z_n(t). \tag{5.67}$$

By taking differential of (5.66), we find that the subsidiary stochastic variables are driven by the set of stochastic differential equations,

$$dZ_0(t) = -\lambda Z_0(t) + d\tilde{W}(t) \tag{5.68}$$

$$dZ_i(t) = \left[i Z_{i-1}(t) - \lambda Z_i(t) \right] dt, \; i = 1,2,...,n \;. \tag{5.69}$$

Note that the stochastic integral equation (5.3) for the spot rate of interest can be written as,

$$r(t) = f(0,t) + \int_0^t \sigma(v,t) \int_v^t \sigma(v,y) dy dv$$
$$+ a Z_0(t) + a_1 Z_1(t) + ... + a_n Z_n(t). \tag{5.70}$$

For ease of notation, we define,

$$D(t) \equiv \frac{\partial}{\partial t}\left[f(0,t) + \int_0^t \sigma(v,t)\int_v^t \sigma(v,y)dydv\right]$$

$$= f_2(0,t) + \int_0^t\left[\sigma_2(v,t)\int_v^t \sigma(v,y)dy + \sigma(v,t)^2\right]dv \qquad (5.71)$$

where, $\sigma_i(.,.)$, $i=1,2$ denotes the partial derivative of $\sigma_i(.,.)$ with respect to the ith argument.

Taking differential of (5.70) we obtain the stochastic differential equation that drives $r(t)$, namely,

$$dr(t) = \begin{bmatrix} D(t) + (a_1 - \lambda a_0)Z_0(t) + ... + \\ (na_n - \lambda a_{n-1})Z_{n-1}(t) - \lambda a_n Z_n(t) \end{bmatrix} dt + a_0 d\tilde{W}(t). \quad (5.72)$$

This equation may be further simplified by noting that, from equation (5.70),

$$aZ_0(t) + a_1 Z_1(t) + ... + a_n Z_n(t)$$
$$= r(t) - f(0,t) - \int_0^t \sigma(v,t)\int_v^t \sigma(v,y)dydv. \qquad (5.73)$$

Therefore,

$$dr(t) = \begin{bmatrix} D(t) - \lambda r(t) + \lambda f(0,t) + \lambda \int_0^t \sigma(v,t)\int_v^t \sigma(v,y)dydv \\ + a_1 Z_0(t) + 2a_2 Z_1(t) + ... + na_n Z_{n-1}(t) \end{bmatrix} dt + a_0 d\tilde{W}(t).$$

$$(5.74)$$

The stochastic dynamic system of the variables $B(t,T), r(t), Z_i(t)$, $i=0,1,...,n-1$ is then driven by the $(n+2)$ dimensional Markovian system given by the equations (5.13) and (5.14) where,

$$J(t) = \begin{bmatrix} -\sigma_B^2(t,T) \\ \frac{\partial}{\partial t}g(t) + \lambda g(t) \\ 0 \\ \cdot \\ 0 \end{bmatrix} \qquad (5.75)$$

where

$$g(t) = f(0,t) + \int_0^t \sigma(v,t) \int_v^t \sigma(v,y) dy dv$$

and,

$$V(t) = \left[\sigma_B(t,T), a_0, 1, 0..0\right]^T .$$ (5.76)

Finally,

$$H = \begin{bmatrix} 0 & 1 & 0 & . & . & . & 0 \\ 0 & -\lambda & a_1 & 2a_2 & . & . & na_n \\ 0 & 0 & -\lambda & 0 & . & . & 0 \\ 0 & 0 & 1 & -\lambda & 0 & . & 0 \\ 0 & 0 & 0 & 2 & -\lambda & . & 0 \\ . & . & . & . & . & . & . \\ 0 & 0 & 0 & 0 & . & n-1 & -\lambda \end{bmatrix} .$$ (5.77)

Appendix 5.3: Proof of Proposition 2

If we now define the subsidiary stochastic variables,

$$Z_i(t) = \int_0^t G(r(u))(t-u)^i e^{-\lambda(t-u)} d\tilde{W}(u)$$ (5.78)

then,

$$\int_0^t G(r(u)) \sigma(u,t) d\tilde{W}(u) = a_0 Z_0(t) + a_1 Z_1(t) .$$ (5.79)

Taking differentials of (5.78) we obtain,

$$dZ_0(t) = -\lambda Z_0(t) + G(r(t)) d\tilde{W}(t)$$ (5.80)

$$dZ_1(t) = \left[Z_0(t) - \lambda Z_1(t)\right] dt .$$ (5.81)

With the help of (5.78), (5.3) can be written as,

$$r(t) = f(0,t) + \int_0^t G(r(v))^2 \sigma(v,t) \int_v^t \sigma(v,y) dy dv$$
$$+ a_0 Z_0(t) + a_1 Z_1(t),$$ (5.82)

which upon taking differentials may be expressed as the stochastic differential equation,

$$dr(t) = D_r(t)dt + a_0 dZ_0(t) + a_1 dZ_1(t) \qquad (5.83)$$

where,

$$D_r(t) = f_2(0,t) + \int_0^t G(r(v))^2 \Sigma(v,t)dv \qquad (5.84)$$

and,

$$\Sigma(v,t) \equiv \sigma_2(v,t)\int_v^t \sigma(v,y)dy + \sigma(v,t)^2. \qquad (5.85)$$

Equation (5.83) with the help of (5.80) and (5.81) becomes,

$$dr(t) = \begin{bmatrix} D_r(t) + \lambda f(0,t) + \\ \lambda \int_0^t G(r(v))^2 \sigma(v,t)\int_v^t \sigma(v,y)dydv - \\ \lambda r(t) + a_1 Z_0(t) \end{bmatrix} dt + a_0 G(r(t))d\tilde{W}(t)$$

$$(5.86)$$

Using equation (5.26) the term $D_r(t)$ and the integral in equation (5.86) can be expressed as,

$$D_r(t) + \lambda \int_0^t G(r(v))^2 \sigma(v,y)\int_v^t \sigma(v,y)dydv$$

$$= f_2(0,t) + \phi_{0,1}(t)\left[\frac{a_1^2}{\lambda^2} + \frac{a_0 a_1}{\lambda}\right] + \phi_{1,1}(t)\left[a_0 a_1 + \frac{a_1^2}{\lambda} + \frac{a_1^2}{\lambda^2} - \frac{a_0 a_1}{\lambda}\right] \qquad (5.87)$$

$$+ \phi_{0,2}(t)\left[a_0^2 - \frac{a_1^2}{\lambda} - \frac{a_0 a_1}{\lambda}\right] + \phi_{1,2}(t)\left[2a_0 a_1 - \frac{a_1^2}{\lambda}\right] + \phi_{2,2}(t)\left[a_1^2\right]$$

where, we define the additional subsidiary variables,

$$\phi_{k,1} = \int_0^t G(r(u))^2 (t-u)^k e^{-\lambda(t-u)}du, \, k = 0,1 \qquad (5.88)$$

$$\phi_{k,2} = \int_0^t G(r(u))^2 (t-u)^k e^{-2\lambda(t-u)}du, \, k = 0,1,2. \qquad (5.89)$$

Equation (5.86), therefore, turns into,

$$dr(t) = \begin{bmatrix} f_2(0,t) + \lambda f(0,t) - \lambda r(t) + a_1 Z_0(t) \\ + \phi_{0,1}(t)\left(\dfrac{a_1^2 + a_0 a_1 \lambda}{\lambda^2}\right) \\ + \phi_{1,1}(t)\left(\dfrac{a_0 a_1 \lambda^2 + a_1^2 \lambda - a_1^2 - a_0 a_1 \lambda}{\lambda^2}\right) \\ + \phi_{0,2}(t)\left(\dfrac{a_0^2 \lambda^2 - a_1^2 - a_0 a_1 \lambda}{\lambda^2}\right) \\ + \phi_{1,2}(t)\left[2a_0 a_1 - \dfrac{a_1^2}{\lambda}\right] \\ + \phi_{2,2}(t)a_1^2 \end{bmatrix} dt + a_0 G(r(t)) d\tilde{W}(t).$$

(5.90)

Taking differentials of (5.88) and (5.89), we find that the subsidiary state variables are determined by the stochastic differential equations,

$$d\phi_{0,1}(t) = \left[G(r(t))^2 - \lambda\phi_{0,1}(t)\right]dt,$$ (5.91)

$$d\phi_{1,1}(t) = \left[\phi_{0,1}(t) - \lambda\phi_{1,1}(t)\right]dt,$$ (5.92)

$$d\phi_{0,2}(t) = \left[G(r(t))^2 - 2\lambda\phi_{0,2}(t)\right]dt,$$ (5.93)

$$d\phi_{1,2}(t) = \left[\phi_{0,2}(t) - 2\lambda\phi_{1,2}(t)\right]dt,$$ (5.94)

$$d\phi_{2,2}(t) = \left[2\phi_{1,2}(t) - 2\lambda\phi_{2,2}(t)\right]dt.$$ (5.95)

Now, grouping the equations (5.25), (5.80) and (5.91) – (5.95), the stochastic dynamical system of equations under consideration can be expressed as the Markovian system (5.27), where,

$$J(r,t) \equiv G(r(t))^2 \begin{bmatrix} -0.5G(r(t))^2 \sigma_B^2(t,T), f_2(0,t) + \lambda f(0,t), 0, \\ G(r(t))^2, 0, G(r(t))^2, 0, 0 \end{bmatrix}^T,$$ (5.96)

$$V(r,t) \equiv \left[G(r(t))\sigma_B(t,T), a_0 G(r(t)), G(r(t)), 0, 0, 0, 0, 0\right]^T,$$ (5.97)

and,

$$H \equiv \begin{bmatrix} 0 & 1 & 0 & 0 & 0 & 0 & 0 & 0 \\ 0 & -\lambda & a_1 & H_{2,4} & H_{2,5} & H_{2,6} & H_{2,7} & H_{2,8} \\ 0 & 0 & -\lambda & 0 & 0 & 0 & 0 & 0 \\ 0 & 0 & 0 & -\lambda & 0 & 0 & 0 & 0 \\ 0 & 0 & 0 & 1 & -\lambda & 0 & 0 & 0 \\ 0 & 0 & 0 & 0 & 0 & -2\lambda & 0 & 0 \\ 0 & 0 & 0 & 0 & 0 & 1 & -2\lambda & 0 \\ 0 & 0 & 0 & 0 & 0 & 0 & 2 & -2\lambda \end{bmatrix}, \qquad (5.98)$$

where,

$$H_{2,4} = \frac{1}{\lambda^2}\left(a_1^2 + a_0 a_1 \lambda\right), \ H_{2,5} = \frac{1}{\lambda^2}\left(a_0 a_1 \lambda^2 + a_1^2 \lambda - a_1^2 - a_0 a_1 \lambda\right)$$

$$H_{2,6} = \frac{1}{\lambda^2}\left(a_0^2 \lambda^2 - a_1^2 - a_0 a_1 \lambda\right), \ H_{2,7} = \frac{1}{\lambda}\left(2a_0 a_1 \lambda^2 - a_1^2\right), \ H_{2,8} = a_1^2.$$

Appendix 5.4: Proof of Proposition 3

The approach so far has been based on the form of the polynomial (5.12) with n=1. The methodology of deriving the Markovian system can be extended to the case of general n. The following steps outline this general case. Let,

$$Z_m(t) = \int_0^t G\left(r(v)\right)(t-v)^m \, d\tilde{W}(v), \ m = 0,1,...,n \qquad (5.99)$$

then,

$$r(t) = f(0,t) + \int_0^t G\left(r(v)\right)^2 \sigma(v,t) \int_v^t \sigma(v,y) dy dv$$
$$+ aZ_0(t) + a_1 Z_1(t) + ... + a_n Z_n(t). \qquad (5.100)$$

Note furthermore that,

$$dZ_0(t) = -\lambda Z_0(t)dt + G\left(r(t)\right)d\tilde{W}(t) \qquad (5.101)$$

whilst for m $\geq$ 1,

$$dZ_m(t) = \left[mZ_{m-1}(t) - \lambda Z_m(t)\right]dt. \qquad (5.102)$$

Taking differential of equation (5.100) we obtain the stochastic differential form for r(t),

$$dr(t) = D_r(t)dt + a_0 dZ_0(t) + a_1 dZ_1(t) + \ldots + a_n dZ_n(t) \quad (5.103)$$

where we define,

$$D_r(t) = f_2(0,t) + \int_0^t G(r(v))^2 \Sigma(v,t)dv \quad (5.104)$$

and,

$$\Sigma(v,t) \equiv \sigma_2(v,t) \int_v^t \sigma(v,y)dy + \sigma(v,t)^2. \quad (5.105)$$

A slight rearrangement yields,

$$dr(t) = \begin{bmatrix} D_r(t) + (a_1 - \lambda a_0)Z_0(t) + \ldots + (a_n - \lambda a_{n-1})Z_{n-1}(t) \\ -\lambda a_n Z_n(t) \end{bmatrix} \quad (5.106)$$
$$\times dt + a_0 G(r(t))d\tilde{W}(t)$$

where, r(t) can be back substituted from equation (5.100) to yield,

$$dr(t) = \begin{bmatrix} D_r(t) + \lambda f(0,t) + \\ \lambda \int_0^t G(r(v))^2 \sigma(v,t)\int_v^t \sigma(v,y)dydv + \\ a_1 Z_0(t) + \ldots + a_n Z_{n-1}(t) - \lambda r(t) \end{bmatrix} dt + a_0 G(r(t))d\tilde{W}(t). \quad (5.107)$$

The only difficulty remaining is the term.

$$\int_0^t G(r(v))^2 \Sigma(v,t)dv \quad (5.108)$$

appearing (5.104). Given the assumed functional form for $\sigma(v,t)$ this last integral can be expressed as a sum of terms involving,

$$\phi_{j,k}(t) = \int_0^t G(r(v))^2 (t-v)^j e^{-\lambda k(t-v)}dv, \ 0 < k \le n+1, \ 0 \le j \le k \quad (5.109)$$

which describes statistical characteristics of the history of the process for r(t). Note that for each k,

$$d\phi_{0,k}(t) = \left[G(r(t))^2 - \lambda k\phi_{0,k}(t) \right]dt \quad (5.110)$$

whilst for $1 \leq j \leq n+1$

$$d\phi_{j,k}(t) = \left[j\phi_{j-1,k}(t) - \lambda k\phi_{j,k}(t) \right] dt . \qquad (5.111)$$

Hence, we again obtain a Markovian system, the order being determined by n. For n=1, the number of ϕ subsidiary variables required is five and in general $0.5(n+2)(n+3)-1$ of the ϕ variables. Also, n of the Z_m subsidiary stochastic variables will be required, m varying from 0 to n-1.

Chapter 6

Filtering and Hedging using Interest Rate Futures

Interest rate futures contracts are important in hedging interest rate exposures by firms and financial institutions. Volatility of the interest rate process is a key determinant how such financial hedges are constructed. Although, historical interest rate data may be used to get an estimate of the volatility, however, it would still be historical in nature. In this chapter we show how to employ HJM (Heath-Jarrow-Morton) framework to infer volatility from traded interest rate futures contracts using stochastic filtering and in particular extended Kalman filter.

Hedging interest rate exposures using interest rate futures contracts requires some knowledge of the volatility function of the interest rates. Use of historical data as well as interest rate options like caps and swaptions to estimate this volatility function have been proposed in the literature. In this chapter the interest rate futures price is modeled within an arbitrage-free framework for a volatility function which includes a stochastic variable, the instantaneous spot interest rate. This makes the whole approach more realistic. The resulting system is expressed in a state space form which is solved using extended Kalman filter. The residual diagnostics indicate suitability of the model and the bootstrap resampling technique is used to obtain small sample properties of the parameters of the volatility function.

6.1 Background Details

In most major markets around the world interest rate futures contracts have enjoyed tremendous growth over the last decade or more. The flexibility of these contracts and lower transaction costs have made these attractive to portfolio managers attempting to change risk return characteristics of their holdings. For example, the duration of a fixed income portfolio can be decreased or increased with the help of suitable interest futures contracts.

Many research articles examined the theoretical pricing models and the difference that result from marking-to-market that are special characteristics of the futures markets. For example, Cox-Ingersoll-Ross (CIR) (1981) finds that the size and the difference between the prices of forward and futures contracts is an increasing function of the covariance between the underlying asset and a stochastic interest rate. On the other hand, Flesaker (1993) shows that, for contract maturities up to six months, interest rate futures prices do not differ significantly from the corresponding forward prices. But, for longer maturities the difference may be significant and depends on the maturity as well as the level of the interest rate volatility. This author also points out that the differences in pricing due to continuous marking-to-market and daily marking-to-market is very small. Flesaker (1993) uses the arbitrage-free framework of Heath-Jarrow-Morton (HJM) (1992) under the assumption of constant volatility of the forward rate.

In a published paper, Jarrow and Turnbull (1994) develop the theoretical framework for delta and gamma hedging with interest rate futures contracts in a multi-factor HJM setting. They also point out the fact that this area is far less developed compared to equity and foreign currency derivatives. To implement the theory developed by Jarrow and Turnbull (1994), it is necessary to compute the parameters of the volatility function that has to be specified for the HJM model. They specify the following form of the volatility function:

$$\sigma_f = a_0 \, e^{-\lambda(T-t)}. \qquad (6.1)$$

where, σ_f is the volatility of the forward rate $f(t,T)$ at time t for T period maturity, a_0 and λ are the parameters of the specification. It is suggested that practitioners normally estimate these parameters by fitting the theoretical models to market prices of such instruments as caps and swaptions. Musiela et al (1992) also suggest an algorithm that uses historical data.

This chapter presents an alternative way to infer these parameters using non-linear filtering algorithm. The procedure relies upon a prior specification of the form of the volatility function in the HJM framework, that is more general than in Flesaker (1993) and differs from the equation (6.1) by a multiplicative factor which is a simple function of the instantaneous spot interest rate, r(t). In this construction, the futures price, with continuous marking-to-market, is represented by a stochastic differential equation system of dimension three. The appropriate discretisation method, the filtering algorithm, and the computational details are also described. The empirical results are presented based on the data on short term interest rate futures contracts from both the Sydney Futures Exchange (SFE) and the Tokyo International Financial Futures Exchange (TIFFE).

6.2 The Futures Price Model in the HJM Framework

The starting point in the HJM, (1992) model of the term structure of interest rates is the stochastic integral equation for the forward rate,

$$f(t,T) = f(0,T) + \int_0^t \alpha(u,T)\,du + \int_0^t \sigma_f(u,T)\,dW(u),\ 0 \le t \le T \quad (6.2)$$

where f(t, T) is the forward rate at time t applicable to time T (>t). The single source of noise $dW(u)$ is the increment of a standard Wiener process generated by a probability measure Q. The functions $\alpha(u,T)$ and $\sigma_f(u,T)$ are the instantaneous drift and volatility functions at time u for the forward rate f(t, T). According to HJM (1992) avoidance of riskless arbitrage opportunity and application of Girsanov's theorem transforms the equation (6.2) to,

$$f(t,T) = f(0,T) + \int_0^t \sigma_f(u,T) \int_u^T \sigma_f(u,y)dy\,du + \int_0^t \sigma_f(u,T)d\tilde{W}(u) \quad (6.3)$$

where $d\tilde{W}(u)$ is the increment of a standard Wiener process generated by an equivalent probability measure. The Radon-Nikodym theorem relates the two probability measures Q and $\tilde{Q}$, but the details are not necessary for the purposes of this chapter. The spot rate process, by definition, is given by $r(t) = f(t,t)$ and thus satisfies the stochastic integral equation,

$$r(t) = f(0,t) + \int_0^t \sigma_r(u,t) \int_u^t \sigma_r(u,y)dy \, du + \int_0^t \sigma_r(u,t) \, d\tilde{W}(u). \quad (6.4)$$

In the HJM (1992) context, the price, $P(t,T)$, of a pure discount bond at time t, paying \$1 at time T is given by,

$$P(t,T) = \exp\left(- \int_t^T f(t,s) \, ds\right), \quad 0 \le t \le T, \quad (6.5)$$

which, on application of Ito's lemma gives the dynamic of the bond price, (Jarrow and Turnbull, (1994)),

$$P(t,T) = \frac{P(0,T)}{P(0,t)} \exp\left[\begin{array}{l} -\int_t^T ds \int_0^t \sigma_r(u,s) \int_u^T \sigma_r(u,v)dvdu - \\ \int_t^T ds \int_0^t \sigma_r(u,s)d\tilde{W}(u) \end{array} \right] \quad (6.6)$$

Let $F(t; n)$ denote the futures price at time $t(\le n)$ of a futures contract which matures at time n. The futures contract is written on a discount instrument which matures at time $T(\ge n)$. When the futures contract matures,

$$F(n;n) = P(n;T)$$

and at time $t(\le n)$

$$F(t;n) = E[P(n,T)|\vartheta_t]$$

where ϑ_t denotes the information at time t.

Using equation (6.6) it follows that,

$$F(t; n) = F(0; n) \exp\left[\begin{array}{l} -\frac{1}{2} \int_0^t du \left(\int_n^T \sigma_r(u,s) \, ds\right)^2 - \\ \int_0^t d\tilde{W}(u) \int_n^T \sigma_r(u,s) \, ds \end{array} \right] \quad (6.7)$$

It is assumed that $\sigma_f(u,t)$ has the following form (equation (6.8)), i.e. the product of the equation (6.1) and a stochastic term dependent on the instantaneous spot interest rate,

$$\sigma_f(t,T) = a_0\, e^{-\lambda(T-t)}\, r(t)^\gamma. \tag{6.8}$$

It is shown in the following paragraphs that in this situation the futures price process can be modeled as a three dimensional stochastic differential equation system. Equation (6.8) is motivated by the empirical literature on the short rate process, in particular, Chan et al (1992) and Brenner, Harjes and Kroner (1996). As we see below the form (6.8) implies a process for $r(t)$ whose diffusion term is proportional to $r(t)^\gamma$ and the cited empirical literature lends support to such a dependence. Furthermore, Chiarella and El-Hassan (1996a, 1997) show that with the volatility function (6.8) the HJM model reduces to the Hull and White (1990) extended Vasicek model when $\gamma = 0$ and extended CIR (1985) model when $\gamma = 0.5$.

In order to express equation (6.7) in differential form, it is formulated as,

$$F(t; n) = F(0; n)\exp\left[-X1 - X2\right]$$

$$X1 = \frac{1}{2}\int_0^t du \left(\int_n^T \sigma_f(u,s)\,ds\right)^2 \tag{6.9}$$

$$X2 = \int_0^t d\tilde{W}(u)\int_n^T \sigma_f(u,s)\,ds$$

and this leads to,

$$dX1 = \frac{1}{2}\left(\int_n^T \sigma_f(t,s)\,ds\right)^2 dt, \tag{6.10}$$

$$dX2 = \left(\int_n^T \sigma_f(t,s)\,ds\right) d\tilde{W}(t). \tag{6.11}$$

Now, let $Y = -X_1 - X_2$, then

$$dY = -\frac{1}{2}\left(\int_n^T \sigma_f(t,s)\,ds\right)^2 dt - \left(\int_n^T \sigma_f(t,s)\,ds\right) d\tilde{W}(t) \tag{6.12}$$

or,

$$dY = \mu_Y dt + \sigma_Y d\tilde{W}(t). \tag{6.13}$$

Rewriting equation (6.7) as, $F(t;n) = F(0;n)\exp(Y)$ then by Ito's lemma,

$$dF(t;n) = \left[F(0;n)\, e^{Y}\mu_Y + F(0;n)\, e^{Y}\frac{1}{2}\sigma_Y^2 \right] dt$$

$$+ F(0;\, n)\, e^{Y}\sigma_Y\, d\tilde{W}(t)$$

which simplifies to (by use of the definition of μ_Y and σ_Y in the equation (6.13)),

$$dF(t;n) = F(t;n)\left(\int_n^T \sigma_r(t,s)ds \right) d\tilde{W}(t). \tag{6.14}$$

The stochastic differential equation (6.14), thus, describes the evolution of the futures price in the arbitrage free framework as a function of the volatility of the forward rate. As it has been assumed that the forward rate volatility is given by the equation (6.8), it implies that the dynamics of the futures price in equation (6.14) is, in fact, driven by the dynamics of the spot interest rate given by the equation (6.4). It can be seen from the equation (6.4) that the spot rate process, in general, is non-Markovian due to the presence of the last integral which represents the accumulated effect of the shocks since t=0. It has been shown in Chapter 5 and also in Bhar and Chiarella (1995a) that this spot rate process can be converted to a Markovian system with the expansion of states for a general class of volatility function. That result is summarized below for the volatility function assumed in this paper:

$$dr(t) = [f_2(0,t) + \lambda\, f(0,t)-\lambda\, r(t) + a_0^2\, \phi(t)]\, dt + a_0 r(t)^{\gamma}d\tilde{W}(t) \tag{6.15}$$

$$d\phi(t) = \left[r(t)^{2\gamma} - 2\lambda\phi(t) \right] dt \tag{6.16}$$

where $f_2(0,t)$ is the partial derivative of $f(0,t)$ with respect to the second argument and the derivation is summarized in the Appendix of this chapter.

The state variable $\phi(t)$ which summarizes the characteristics of the path history of the instantaneous spot rate process is not readily observable. It will be further assumed that r(t) itself is not directly observable. Depending on the application, however, some proxy variable e.g. overnight rate or 30-day inter-bank rate may represent r(t). The stochastic differential equations (6.14) - (6.16), therefore, jointly determine the futures price in this model. It should, however, be stressed that the distribution of r(t) in this model is under the equivalent martingale probability measure.

It should be pointed out that similar reduction to Markovian system has been found by Carverhill (1994) and Ritchken and Sankarasubramanian (RS) (1995) as discussed in Chapter 5. It has also been shown in Bhar and Chiarella (1995a) that the RS (1995) system can be obtained as a special case of their approach.

6.3 Non-Linear Filter for Futures Price System

With the volatility function given in equation (8), the futures price can be expressed as a Markovian stochastic differential system as:

$$dS(t) = [J(t) + H(t) S(t)] dt + V(t) d\tilde{W}(t), \qquad (6.17)$$

$$S(t) = [F(t, n), r(t), \phi(t)]', \qquad (6.18)$$

$$V(t) = \left[F(t;n) a_0 r(t)^\gamma \int_n^T e^{-\lambda(s-t)} ds, \ a_0 r(t)^\gamma, \ 0 \right]', \qquad (6.19)$$

$$J(t) = [0, \ f_2(0,t) + \lambda f(0,t), \ 0]', \qquad (6.20)$$

and,

$$H(t) = \begin{bmatrix} 0 & 0 & 0 \\ 0 & -\lambda & a_0^2 \\ 0 & r(t)^{2\gamma-1} & -2\lambda \end{bmatrix}. \qquad (6.21)$$

Since the only element of state vector considered observable is $F(t;n)$, the observation vector (a scalar in this case) is introduced as,

$$Y(t) = C \times S(t) + e(t), \ C = [1, 0, 0]. \qquad (6.22)$$

The observation error is characterized by $\varepsilon(t) \sim N(0, \sigma_\varepsilon^2)$. It is further assumed that the error sequences in the equation (6.17) and (6.22) are independent. The observation error is believed to be introduced by the presence of spread between open-close or high-low quotes. As the observed variable $F(t;n)$ entering the system through the measurement equation (6.22), the approach to solving this problem is based on non-linear filtering. For ease of exposition the equation (6.17) is expressed as,

$$dS(t) = F(S(t); \theta) \, dt + V(S(t); \theta) \, d\tilde{W}(t), \ \theta = [a_0, \lambda, \gamma]. \quad (6.23)$$

In general F(.) and V(.) will be non-linear in both the state variables as well as the parameters. Estimation of the parameter vector θ will involve some form of discretisation from which the conditional moments over successive time intervals can be calculated. Three different approaches have been described in Bhar and Chiarella (1997) and also refer to Chapter 1. In this section only the Milstein scheme is adopted. The Milstein scheme which is an order 1.0 strong Taylor scheme (see Kloeden and Platen (1992) for details) discretises the stochastic differential equation (6.23) within the interval $\delta_k = t_{k+1} - t_k$, as

$$S_{k+1} = \left[S_k + F(S_k; \theta) \delta_k \right] + V(S_k; \theta) \sqrt{\delta_k} \ \zeta$$
$$+ \frac{1}{2} \dot{V}(S_k; \theta) V(S_k; \theta) \delta_k \left[\left(\tilde{\zeta}_k \right)^2 - 1 \right], \ \tilde{\zeta}_k \sim N(0,1). \quad (6.24)$$

Strong schemes are also required to generate path wise approximations for studying estimators of statistical parameters as recommended by Kloeden, Schurz, Platen and Sorensen (1992).

The derivative in the third term $(\dot{V})$ of the equation (6.24) represents the matrix of derivatives of the elements of the vector V with respect to the elements of the state vector, S. Using V^i to denote the ith element of the vector V, this derivative is evaluated as,

$$\begin{bmatrix} \dfrac{\partial V^1}{\partial F} & \dfrac{\partial V^1}{\partial r} & \dfrac{\partial V^1}{\partial \phi} \\[2mm] \dfrac{\partial V^2}{\partial F} & \dfrac{\partial V^2}{\partial r} & \dfrac{\partial V^2}{\partial \phi} \\[2mm] \dfrac{\partial V^3}{\partial F} & \dfrac{\partial V^3}{\partial r} & \dfrac{\partial V^3}{\partial \phi} \end{bmatrix} = \begin{bmatrix} \dfrac{a_0 r(t)^\gamma \eta}{\lambda} & \dfrac{F(t;n) a_0 \gamma \, r(t)^{\gamma-1} \eta}{\lambda} & 0 \\[2mm] 0 & a_0 \gamma \, r(t)^{\gamma-1} & 0 \\[2mm] 0 & 0 & 0 \end{bmatrix} \quad (6.25)$$

where,

$$\eta = \left[e^{-\lambda(n-t)} - e^{-\lambda(T-t)} \right].$$

Examining the equation (6.24) the first two conditional moments can be written as,

$$E\left(S_{k+1}\big|S_k\right) = S_k + F(S_k;\theta)\delta_k, \tag{6.26}$$

$$\begin{aligned}
\text{Cov}\left(S_{k+1}\big|S_k\right) &= V(S_k;\theta)V'(S_k;\theta)\delta_k \\
&+ \frac{1}{2}(\dot{V}VV' + VV'\dot{V}')\delta_k^{3/2}E(\tilde{\zeta}^3) \\
&+ \frac{1}{4}\dot{V}VV'\dot{V}\delta_k^2(E(\tilde{\zeta}^4)-1).
\end{aligned} \tag{6.27}$$

From equation (26) the best forecast of S at t_{k+1} made at t_k (knowing Y_k at t_k) is,

$$\hat{S}_{k+1|k} = S_k + F(S_k;\theta)\delta_k, \tag{6.28}$$

and the best forecast of variance of S_{k+1} is,

$$P_{k+1\,|\,k} = P_{k\,|\,k} + Q_{k+1} \tag{6.29}$$

where, Q_{k+1} is given by (6.27).

Under the assumption of normal distribution as incorporated in the equations (6.26) and (6.27), the transition probability density function for the state vector S_k to S_{k+1} can be written for a given set of observations with the help of the updating equations for Kalman filter as described in Chapter 1. Tanizaki (1993) is also a good reference.

To estimate the parameter vector θ of the volatility function of the forward rates, the likelihood function $L(\theta)$, formed following the algorithm in Chapter 1 can be maximized using a suitable numerical optimization procedure. Maximizing L with respect to θ yields consistent and asymptotically efficient estimators $\hat{\theta}$ (see Lo (1988)). That is (T being the number of observations),

$$\text{p}\lim_{T\to\infty}\hat{\theta} = \theta, \quad \sqrt{T}\left(\hat{\theta}-\theta\right) \sim N\left(0, I^{-1}(\theta)\right)$$

where the asymptotic covariance matrix $I^{-1}(\theta)$ is given by,

$$I(\theta) = \lim_{T\to\infty}\frac{1}{T}\sum_{R=1}^{\infty}E\left[\frac{\partial^2 \log L}{\partial\theta\partial\theta'}\right].$$

6.4 Data Used in Empirical Study

The model developed in the previous section is applied to short-term interest rate futures contracts traded on the Sydney Futures Exchange (SFE) and on the Tokyo International Financial Futures Exchange (TIFFE). The SFE contracts are on 90-day bank bills for Australian dollar denominated borrowing/lending and the TIFFE contracts are on 3-month Euroyen deposits. Both contracts are deliverable in March, June, September and December. The 90-day bank bill futures contracts mature on second Fridays of the delivery months and the 3-month Euroyen contracts mature on third Wednesdays of the delivery months. Although trading of these contracts five or six quarters ahead are quite common, only the near quarter month contracts are included in the analysis in this paper. Also while nearly a year before the delivery dates these contracts are listed in both these two exchanges, the estimation has been carried out using only the last three months of the trading data. This represents the most active part of trading during the lives of these contracts. Therefore, the average number of observations per contract is 62. The period covered by the data starts from September 1989 contracts and ends with the March 1994 contracts.

The 90-day bank bill contracts trading on the SFE are quoted in terms of an index number obtained from (100 - yield to maturity). Therefore, the futures price per dollar of face value is given by

$$\frac{1}{\left(1 + \text{yield} \times \dfrac{90}{365}\right)}$$

where yield is expressed as a fraction.

In case of the 3-month Euroyen contracts, trading on the TIFFE, the quoted price is an index given by (100 - discount). Thus, the futures price per 100 units of face value is given by

$$\left[100 - (100 - \text{discount}) \times \frac{90}{360}\right].$$

6.5 Empirical Results

The functional form of the volatility of the forward rate given by the equation (6.8) includes a variable γ. The empirical results presented here has been obtained by setting $\gamma = 0.5$. The objective of the estimation procedure is therefore to obtain the parameter vector, $\theta \equiv \begin{bmatrix} a_0 & \lambda \end{bmatrix}'$. Also, σ_ε^2 is given a value of 0.00001 based upon an analysis of the spread between the open-close prices of the futures contracts during the time period examined.

The estimation process also requires specification of the initial forward rate function f(0,T). This is done by using a polynomial fitting and is described in detail in Bhar and Hunt (1993). The initial value of the spot rate is obtained from f(0,T) by setting T=0.

Table 6.1 reports the estimated coefficients together with the values of the log likelihood function for the 19 90-day bank bill futures contracts examined from the SFE. Table 6.2 similarly reports the results for the 19 3-month Euroyen futures contracts from the TIFFE. The standard errors reported in these Tables should be treated with caution due to small sample size.

The bootstrap re-sampling approach to non-parametric confidence interval estimation is discussed below. These Tables also include the sample statistics of the implied instantaneous spot interest rates. These statistics are simply the mean and standard deviation of the time path for r(t) generated by the state equations with the estimated parameter vector $\theta \equiv \begin{bmatrix} a_0 & \lambda \end{bmatrix}'$. The implied mean rates correspond well with the short-term rates in existence around the period in question. This is indicative of reasonable fit of the model with the estimated parameters with γ set to 0.5. However, the question of the value of γ that fits the data best has not been pursued in this analysis.

Table 6.3 gives the residual diagnostics for both sets of data. It is clear that for most of the contracts the hypothesis of serial correlations can be rejected at usual level of significance. This is further proof of statistical significance of the estimated parameters. Regarding normality in the residual, the TIFFE data perform better than the SFE data. This is evident from the Jarque-Bera statistics in Table 6.3. This probably suggests that the value of $\gamma = 0.5$ is not appropriate for the SFE data set.

In order to be able to comment about the statistical confidence interval of the estimated parameters, bootstrap distribution of these parameters is generated using the March 1994 contracts for both the SFE and the TIFFE data. The steps in the bootstrap procedure are as follows:

i.   Using the original data set, the parameters are estimated as explained in the text by maximizing the prediction error decomposition form of the likelihood function.

ii.   The standardized prediction error vector ($e_k^s(\theta) = e_k(\theta) v_k^{-0.5}$) is obtained. This in turn generates the innovations of the state transition equation ($\tilde{\zeta}_{\flat k}(\theta) = K_k e_k^s(\theta)$). More information regarding this can be found in Anderson and Moore (1979, page 231). Here, prediction error and its variance have the same definition as in Chapter 1. K_k is the Kalman gain at the kth observation, also defined in Chapter 1.

iii. By sampling $\tilde{\zeta}_{\flat k}(\theta)$, and $e_k^s(\theta)$ using uniformly distributed random numbers one instance of bootstrap innovations for the state transition equation and the measurement equation is obtained.

iv.   The innovations from the step (iii) are used for a complete recursion through the equations (6.22) and (6.24) in place of the original error sequences. This generates one instance of bootstrap observations $Y^B(k)$. The initial conditions are kept same as in the original maximum likelihood estimation in step (i).

v.   The steps (iii) and (iv) are repeated to obtain 1000 bootstrap replications. For each such replication the maximum likelihood estimation is carried out and a histogram of the distribution of the estimated parameters and sample statistics are obtained.

In the Figures 6.2 and 6.3 the distribution of these parameters for the SFE data are shown. The quantiles of the distributions are also given as a measure for the confidence interval. Similarly, the Figures 6.6 and 6.7 relate to the TIFFE data. In case of SFE data the estimated a_0 lies within 1.42 standard deviation of the mean of the bootstrap distribution, where as λ lies within 0.41 standard deviation of the bootstrap distribution.

Similarly, in case of TIFFE data the estimated a_0 lies within 0.96 standard deviation of the mean of the bootstrap distribution where as λ lies within 0.48 standard deviation of the bootstrap distribution. On this basis the statistical significance of the estimates can be claimed.

However, these bootstrap distributions are subject to median bias as explained in Efron (1987). This bias problem can be overcome by considering, amongst a number of other methods, a distribution of the pivotal quantity e. g. bootstrap-t.

$$\left(a_0^B - a_0\right)/SE(a_0^B), \quad \left(\lambda^B - \lambda\right)/SE(\lambda^B)$$

These quantities are depicted in the Figures 6.4-6.5 and in the Figures 6.8-6.9. The superscript B denotes bootstrap estimate and SE stands for standard error. This pivotal quantity (bootstrap-t) distribution provides improved estimates of confidence interval compared to ordinary bootstrap distribution where median bias may be present.

This pivotal quantity for the 90-day bank bill futures data for a_0, λ are -1.42 and -0.41 respectively. Referring to the Figures 6.4 and 6.5 it is clear that both these values are within the 95% confidence intervals. Similarly, these pivotal quantities for a_0, λ for the 3-month Euroyen futures data are -0.96 and -0.48 respectively. Both these values also lie within the 95% confidence interval as can be seen from the Figures 6.8 and 6.9. The bootstrap distributions also clearly indicate that only a certain combinations of the parameters a_0 and λ have high probability of occurring. This provides further support to the appropriateness of the estimation method developed here.

Implementation of a recursive algorithm like Kalman filter requires careful specification of the prior values, particularly, the prior covariance matrix of the state vector. The error in the estimate of the state vector is not only dependent on the data but also on P_0 since $P_{k+1|k}$ is partly determined by P_0. It is therefore interesting to determine the effect of the change in P_0 on subsequent $P_{k+1|k}$. This is achieved by defining the quantity sensitivity in the Figure 6.1. It can be seen from this figure that the effect of a change in the prior covariance specification settles down within one iteration. This indicates that the system is not overly sensitive to the prior covariance specification. However, in the absence of sufficient knowledge, it has been suggested (see Harvey (1989), page

121) that $P_0 = w*I$ be specified, where I is the identity matrix and w is large scalar, e.g. 10000. In this way the filter is initialized with diffuse prior information.

Before concluding this section a few comments regarding the numerical procedure adopted in this paper are in order. Maximization of the log likelihood function is carried out by the OPTMUM routine of the Gauss™ package. For the type of data sets used in this study convergence time is only a few seconds per futures contract. However, like any non-linear optimization the convergence time is sensitive to the appropriate initial value specification. Results presented here are robust to different initial value specifications.

6.6 Concluding Remarks

This chapter presents a technique to estimate the parameters of the volatility function of the interest rate process determining the prices of interest rate futures in an arbitrage-free framework. The volatility function includes a stochastic variable represented by the instantaneous spot interest rate as well as a deterministic function of time. It is shown that the resulting system of stochastic differential equations can be represented in the state space form which can be estimated using non-linear filter.

The technique has been applied to short-term interest rate futures contracts from both the Sydney Future Exchange and the Tokyo International Financial Futures Exchange. The effectiveness of the estimation method has been analyzed using the residual series and are found to have the desirable properties. The significance of the estimated parameters has also been established using non-parametric bootstrap simulations.

The question of whether the estimated volatility function consistently prices other derivative securities has not been addressed in this paper. However, some of these issues following from the estimation method presented here have been addressed in Chiarella and El-Hassan (1996b, 1997).

An extension of this methodology developed here is to apply it to interest rate futures contracts of more than one delivery dates. For example, the parameters of the volatility function may be estimated using bank bill futures contracts deliverable over two consecutive quarters. The same concept is applicable to the Euroyen futures contracts as well.

Appendix 6.1

Using the definition forward rate volatility in equation (6.8), the stochastic integral equation (6.4) for r(t) may be expressed as,

$$r(t) = f(0,t) + \int_0^t a_0^2 \, r(u)^{2\gamma} \int_u^t e^{-\lambda[(t-u)+(y-u)]} dy \, du$$

$$+ \int_0^t a_0 r(u)^{\gamma} e^{-\lambda(t-u)} d\tilde{W}(u),$$

(6.30)

$$r(t) = f(0,t) + \frac{a_0^2}{\lambda} \int_0^t r(u)^{2\gamma} e^{-\lambda(t-u)} du - \frac{a_0^2}{\lambda} \int_0^t r(u)^{2\gamma} e^{-2\lambda(t-u)} du$$

$$+ a_0 \int_0^t r(u)^{\gamma} e^{-\lambda(t-u)} d\tilde{W}(u).$$

(6.31)

Defining the subsidiary variables,

$$\phi_{0,1}(t) = \int_0^t r(u)^{2\gamma} e^{-\lambda(t-u)} du,$$

(6.32)

$$\phi_{0,2}(t) = \int_0^t r(u)^{2\gamma} e^{-2\lambda(t-u)} du,$$

(6.33)

$$z_0(t) = \int_0^t r(u)^{\gamma} e^{-\lambda(t-u)} d\tilde{W}(u).$$

(6.34)

With the help of (6.32) - (6.34), (6.31) can be written as,

$$r(t) = f(0,t) + \frac{a_0^2}{\lambda} \phi_{0,1}(t) - \frac{a_0^2}{\lambda} \phi_{0,2}(t) + a_0 z_0(t).$$

(6.35)

Taking differentials of (6.32) - (6.34),

$$d\phi_{0,1}(t) = [r(t)^{2\gamma} - \lambda \phi_{0,1}(t)]dt,$$

(6.36)

$$d\phi_{0,2}(t) = [r(t)^{2\gamma} - 2\lambda \phi_{0,2}(t)]dt,$$

(6.37)

$$dz_0(t) = -\lambda z_0(t) + r(t)^{\gamma} d\tilde{W}(t).$$

(6.38)

Hence, equation (6.35) can now be written in differential form using (6.36) - (6.38),

$$dr(t) = [f_2(0,t) - a_0^2 \phi_{0,1}(t) + 2a_0^2 \phi_{0,2}(t) - \lambda a_0 z_0(t)]dt + a_0 r(t)^\gamma d\tilde{W}(t).$$

$$(6.39)$$

However from (6.35),

$$\lambda r(t) - \lambda f(0,t) + a_0^2 \phi_{0,2}(t) = a_0^2 \phi_{0,1}(t) + \lambda a_0 z_0(t), \qquad (6.40)$$

which when substituted in (6.39) gives,

$$dr(t) = [f_2(0,t) + \lambda f(0,t) + a_0^2 \phi_{0,2}(t) - \lambda r(t)]dt + a_0 r(t)^\gamma d\tilde{W}(t). \quad (6.41)$$

Table 6.1 Estimation SFE Bank Bill Futures.

Contract	a_0	λ	Log Likelihood.	Implied Spot Rate Mean	Std. Dev.
Sep. 1989	0.00187	-11.57442	-431.69	0.1269	0.0633
	(0.00001)	(0.26943)			
Dec. 1989	0.00889	-6.29805	-448.75	0.0760	0.0158
	(0.00001)	(0.28785)			
Mar. 1990	0.01573	-5.85906	-392.61	0.0773	0.0159
	(0.00001)	(0.30718)			
Jun. 1990	0.00056	-13.84868	-446.92	0.0836	0.0352
	(0.00040)	(1.91649)			
Sep. 1990	0.13102	4.33495	-473.78	0.0680	0.0091
	(0.05313)	(1.89958)			
Dec. 1990	0.08770	1.66162	-429.31	0.0851	0.0113
	(0.01101)	(0.53510)			
Mar. 1991	0.06770	2.14568	-385.95	0.0791	0.0118
	(0.15908)	(9.50070)			
Jun. 1991	0.07987	3.42275	-470.58	0.0811	0.0117
	(0.01842)	(0.86161)			
Sep. 1991	0.25339	6.94233	-441.16	0.0757	0.0143
	(0.01487)	(0.40827)			
Dec. 1991	0.14790	5.15174	-428.05	0.0979	0.0144
	(0.01723)	(0.59136)			
Mar. 1992	0.00252	-9.88192	-423.41	0.0643	0.0143
	(0.00206)	(2.58737)			
Jun. 1992	0.01996	-2.20689	-451.78	0.0529	0.0073
	(0.01733)	(3.29487)			
Sep. 1992	0.14419	4.30487	-451.23	0.0520	0.0097
	(0.02909)	(1.12774)			
Dec. 1992	0.00355	-8.87842	-464.97	0.0606	0.0117
	(0.00166)	(1.48274)			
Mar. 1993	0.04240	0.88222	-429.60	0.0491	0.0064
	(0.00862)	(0.78970)			
Jun. 1993	0.01935	-2.38675	-460.29	0.0503	0.0065
	(0.00527)	(0.97900)			
Sep. 1993	0.03015	0.07056	-478.99	0.0504	0.0065
	(0.00786)	(1.00500)			
Dec. 1993	0.01298	-3.11939	-488.15	0.0509	0.0065
	(0.00516)	(1.42447)			
Mar. 1994	0.01398	0.33134	-496.79	0.0577	0.0076
	(0.00038)	(0.39318)			

Numbers in parentheses are standard errors.

Table 6.2 Estimation EuroYen Futures.

Contract	a_0	λ	Log Likelihood.	Implied Spot Rate Mean	Std. Dev.
Sep. 1989	0.02554 (0.00009)	0.15839 (0.40451)	-426.99	0.0486	0.0068
Dec. 1989	0.00791 (0.00001)	-7.56936 (0.29568)	-415.26	0.0556	0.0086
Mar. 1990	0.01604 (0.00003)	-3.39943 (0.33377)	-433.12	0.0475	0.0064
Jun. 1990	0.00334 (0.00005)	-7.62007 (0.30590)	-457.84	0.0577	0.0098
Sep. 1990	0.06639 (0.00005)	2.23176 (0.33775)	-479.35	0.0390	0.0075
Dec. 1990	0.00689 (0.00003)	-6.05398 (0.31957)	-442.80	0.0548	0.0080
Mar. 1991	0.02231 (0.00006)	-3.21829 (0.34614)	-408.67	0.0486	0.0065
Jun. 1991	0.02215 (0.00006)	-1.46132 (0.34614)	-451.60	0.0469	0.0063
Sep. 1991	0.03269 (0.00005)	-0.54846 (0.34715)	-465.08	0.0500	0.0064
Dec. 1991	0.00397 (0.00005)	-6.94571 (0.30688)	-467.26	0.0558	0.0085
Mar. 1992	0.03274 (0.00004)	-0.99355 (0.37844)	-417.11	0.0520	0.0070
Jun. 1992	0.01012 (0.00013)	-2.06550 (0.35077)	-485.44	0.0485	0.0064
Sep. 1992	0.06612 (0.00004)	2.11638 (0.38723)	-460.90	0.0527	0.0070
Dec. 1992	0.01176 (0.00013)	-1.95498 (0.34432)	-486.95	0.0481	0.0063
Mar. 1993	0.00433 (0.00045)	-4.98060 (0.04786)	-485.76	0.0530	0.0073
Jun. 1993	0.00267 (0.00028)	-6.33375 (0.04564)	-480.14	0.0540	0.0079
Sep. 1993	0.03741 (0.00011)	2.16001 (0.39089)	-505.41	0.0495	0.0064
Dec. 1993	0.01307 (0.00009)	-2.55519 (0.34764)	-453.77	0.0516	0.0069
Mar. 1994	0.04941 (0.00005)	0.61786 (0.36037)	-436.10	0.0449	0.0007

Numbers in parentheses are standard errors.

Table 6.3 Residual Diagnostics.

Contract	SFE Bank Bill Futures		EuroYen Futures	
	Q(5)	JB	Q(5)	JB
Sep. 1989	5.03 (0.412)	3.55 (0.169)	13.49 (0.019)	10.02 (0.007)
Dec. 1989	4.29 (0.509)	1.68 (0.434)	5.09 (0.404)	20.98 (0.000)
Mar. 1990	3.75 (0.586)	28.94 (0.000)	7.15 (0.209)	2.72 (0.256)
Jun. 1990	3.55 (0.615)	6.35 (0.049)	1.45 (0.917)	1.10 (0.575)
Sep. 1990	2.65 (0.753)	51.37 (0.000)	5.33 (0.377)	4.55 (0.102)
Dec. 1990	1.98 (0.851)	24.60 (0.000)	6.84 (0.232)	22.65 (0.000)
Mar. 1991	2.88 (0.718)	10.27 (0.006)	4.46 (0.484)	56.96 (0.000)
Jun. 1991	5.24 (0.387)	11.67 (0.001)	2.43 (0.786)	11.83 (0.003)
Sep. 1991	6.35 (0.273)	59.37 (0.000)	2.48 (0.778)	11.23 (0.004)
Dec. 1991	7.15 (0.209)	17.26 (0.000)	4.43 (0.489)	4.50 (0.105)
Mar. 1992	7.34 (0.196)	3.42 (0.181)	8.88 (0.114)	0.28 (0.868)
Jun. 1992	22.28 (0.000)	31.45 (0.000)	1.13 (0.951)	0.13 (0.935)
Sep. 1992	6.43 (0.266)	22.10 (0.000)	2.61 (0.759)	3.67 (0.159)
Dec. 1992	3.61 (0.607)	19.86 (0.000)	2.56 (0.767)	1.24 (0.537)
Mar. 1993	2.31 (0.805)	16.46 (0.000)	3.39 (0.639)	3.20 (0.201)
Jun. 1993	2.39 (0.793)	58.30 (0.000)	5.46 (0.363)	10.91 (0.004)
Sep. 1993	1.41 (0.923)	2.09 (0.350)	8.33 (0.139)	0.55 (0.761)
Dec. 1993	2.11 (0.833)	19.05 (0.000)	10.91 (0.053)	65.05 (0.000)
Mar. 1994	1.42 (0.922)	1.75 (0.417)	2.17 (0.825)	2.25 (0.324)

Numbers in parentheses are p-values. Q(5) represents the Ljung-Box Q-statistic of lag 5 for residual correlations. Under the null hypothesis of no serial correlations this statistic has a Chi-square distribution with 5 degrees of freedom. JB represents Jarque-Bera statistic for test of normality and it has a Chi-square distribution with two degrees of freedom. For p-values less than 0.010, the corresponding null hypothesis can be rejected at the 0.010 level of significance.

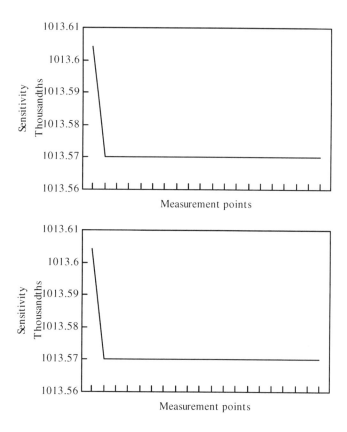

Figure 6.1: Extended Kalman Filter: Error Sensitivity.

Filter performance analyzed using the March 1994 90-day Bank Bill futures contracts (upper panel) and 3-month Euro yen futures contracts (lower panel). In both cases $\gamma = 0.5$. Filter sensitivity is almost identical with both data sets. Sensitivity is measured as the ratio $\left\|\delta P_{k+1|k}\right\|/\left\|\delta P_0\right\|$ where $\|A\| \triangleq \sum a_{ij}^2$. P_0 , $P_{k+1|k}$ are prior covariance and forecast covariance of the state vector at k+1 based on the information at k, respectively. δP_0 refers to the two cases where P_0 is specified as I*100 and I*10000, where I is the identity matrix.

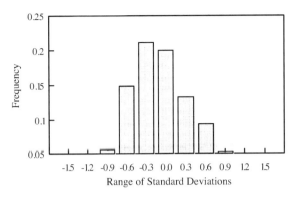

Figure 6.2 Bootstrap Sampling Distribution (a_0).

Frequency distribution of the parameter from 1000 bootstrap re-sampling using March 1994 90-day bank bill futures data. Sample mean and standard deviation are 0.04983, 0.02530 respectively.

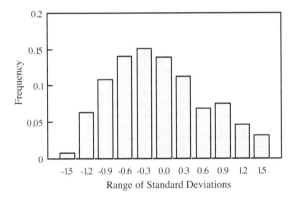

Figure 6.3 Bootstrap Sampling Distribution (λ).

Frequency distribution of the parameter from 1000 bootstrap re-sampling using March 1994 90-day bank bill futures data. Sample mean and standard deviation are 1.18781, 2.09292 respectively.

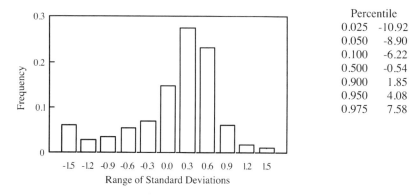

Figure 6.4 Bootstrap Sampling Distribution (Pivotal Quantity for a_0).

Frequency distribution of the pivotal quantity defined in the text from 1000 bootstrap re-sampling using March 1994 90-day bank bill futures data.

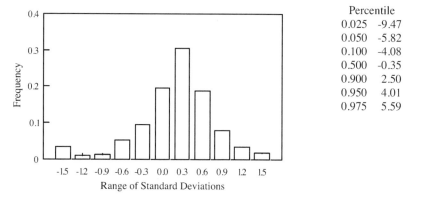

Figure 6.5 Bootstrap Sampling Distribution (Pivotal Quantity for λ).

Frequency distribution of the pivotal quantity defined in the text from 1000 bootstrap re-sampling using March 1994 90-day bank bill futures data.

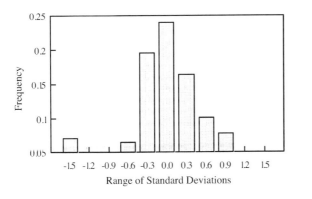

Percentile	
0.025	-0.09099
0.050	-0.06727
0.100	0.08015
0.500	0.14820
0.900	0.26931
0.950	0.33377
0.975	0.40010

Figure 6.6 Bootstrap Sampling Distribution (a_0).

Frequency distribution of the parameter from 1000 bootstrap re-sampling using March 1994 3-month Euroyen futures data. Sample mean and standard deviation are 0.15667, 0.11135 respectively.

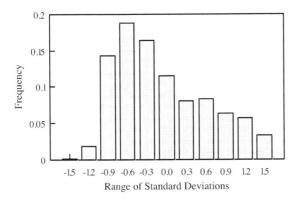

Percentile	
0.025	-0.96854
0.050	-0.82212
0.100	-0.56603
0.500	1.05958
0.900	4.88134
0.950	6.17853
0.975	7.48682

Figure 6.7 Bootstrap Sampling Distribution (λ).

Frequency distribution of the parameter from 1000 bootstrap re-sampling using March 1994 3-month Euroyen futures data. Sample mean and standard deviation are 1.71662, 2.26695 respectively.

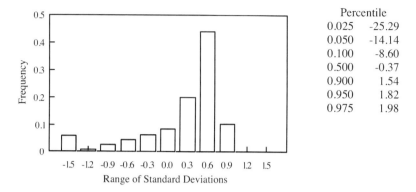

Figure 6.8 Bootstrap Sampling Distribution (Pivotal Quantity for a_0).

Frequency distribution of the pivotal quantity defined in the text from 1000 bootstrap re-sampling using March 1994 3-month Euroyen futures data.

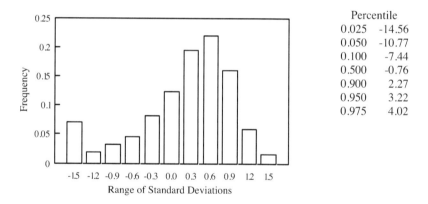

Figure 6.9 Bootstrap Sampling Distribution (Pivotal Quantity for λ).

Frequency distribution of the pivotal quantity defined in the text from 1000 bootstrap re-sampling using March 1994 3-month Euroyen futures data.

Chapter 7

A Multifactor Model of Credit Spreads

We represent credit spreads across ratings as a function of common unobservable factors of the Vasicek form. Using a state-space approach we estimate the factors, their process parameters, and the exposure of each observed credit spread series to each factor. We find that most of the systematic variation across credit spreads is captured by three factors. The factors are closely related to the implied volatility index (VIX), the long bond rate, and S&P500 returns, supporting the predictions of structural models of default at an aggregate level. By making no prior assumption about the determinants of yield spread dynamics, this chapter provides an original and independent test of theory. The results also contribute to the current debate about the role of liquidity in corporate yield spreads. While recent empirical literature shows that the level and time-variation in corporate yield spreads is driven primarily by a systematic liquidity risk factor, this analysis finds that the three most important drivers of yield spread levels relate to macroeconomic variables. This suggests that if credit spread levels do contain a large liquidity premium, the time variation of this premium is likely driven by the same factors as default risk.

7.1 Background and Related Research

The theoretical link between credit spreads and market variables is established by structural models of default. Models such as Merton (1974) and Longstaff and Schwartz (1995) are based on the economic definition of default as the event where a firm's value falls below the face value of its outstanding debt. The unobservable value of the firm is assumed to follow Brownian motion under the assumption of risk-neutrality, allowing the calculation of default probabilities and an endogenous recovery rate.

Credit spreads are attributed entirely to the risk-neutral expected default loss, which is positively related to firm leverage and volatility in the firm value. An increase in the firm value through positive equity performance has the effect of reducing leverage and credit spreads. Under the assumption of risk-neutrality the firm value process has a drift rate equal to the risk-free rate. The models predict that an increase in treasury yields increases the drift of the firm value process, leading to lower credit spreads.

In practice, structural models tend to underestimate short-term credit spreads. The use of smooth processes to represent the firm value may exclude the possibility of default by high grade issuers in the short term, which is inconsistent with the observed role of surprise in credit markets. In contrast, reduced-form models are flexible enough to empirically fit the term structure of credit spreads, but they do not provide an economic interpretation of default. Reduced form models such as Jarrow and Turnbull (1995) and Duffie and Singleton (1999), define exogenous stochastic processes for the arrival time of default and exogenous recovery rates.

An additional class of models combines the advantages of both structural and reduced-form approaches by incorporating exogenous effects such as jump-diffusions (Zhou, 1997) in the firm value process to allow for surprise default. An empirical overview of structural models by Eom, Helwege and Huang (2004) reveals that existing models cannot simultaneously fit both high-grade and low-grade bond spreads. They conclude that more accurate models would need to correct the common tendency to overstate the credit spreads of firms with high leverage or volatility while at the same time understating the spreads of high-grade bonds.

7.2 Variables Influencing Changes in Credit Spreads

To the extent that credit spreads reflect expectations on future default and recovery, we would expect aggregate credit spread indices to vary with macroeconomic variables such as interest rates, stock market returns and market volatility. In general, low-grade bond spreads are observed to be closely related to equity market factors (Huang and Kong, 2003) while high-grade bonds are more responsive to treasury yields. Kwan (1996) finds that individual firm yield spread changes are negatively related to both contemporaneous and lagged equity returns of the same firm. On the other hand, lagged yield spread changes do not help explain current equity returns. Campbell and Taksler (2003) show credit spreads to be positively related to the market average of firm-level equity return volatility, and the increase in market and firm volatility documented by Campbell et al. (2001) is consistent with the steady rise in credit spreads throughout the 1990s.

A negative relationship between investment-grade spreads and treasury yields is estimated by Longstaff and Schwartz (1995), while Duffee (1998) finds that the negative relationship is strongest for callable bonds. Collin-Dufresne et al (2001) show that credit spreads have increasingly negative sensitivities to interest rates as ratings decline across both investment and non investment grade bonds. In a study of only high-yield bonds, Fridson and Jonsson (1995) find no significant relationship between credit spreads and treasury rates, which is more consistent with the idea that low-grade bonds are far more responsive to equity variables than interest rate variables.

While there is strong empirical evidence of a negative relationship between investment-grade spreads and treasury yields, there is no consensus on its economic causes. Intuitively, lower yields should lead to narrower yield spreads through lower borrowing costs that increase the probability of survival. However, falling treasury yields, particularly in the shorter maturities, also tend to be a feature of recessionary periods when default risk rises and central banks typically lower short-term rates. One recent example of this is the sub-prime crisis, beginning in August 2007, during which short-term treasury yields declined to historical lows while credit spreads widened to historical highs. Duffee (1998) concludes that despite the links of both treasury yields and corporate bond spreads to future variations in aggregate output, it is not obvious that these links explain their observed negative relationship, or that yield spreads are determined by credit quality.

To link credit spreads to interest rates and expected aggregate output, the empirical literature has also focused on the slope of the treasury curve, defined as the spread between long-term and short-term yields and often used as a barometer of future economic conditions. Estrella and Hardouvelis (1991) associate a positive slope of the yield curve with future increases in real economic activity, so that an increase (decrease) in the slope of the yield curve should indicate a lower (higher) probability of a recession, in turn reflected in lower (higher) credit spreads. This idea is supported by the findings of Papageorgiou and Skinner (2006) that investment-grade credit spreads are negatively related to changes in both the level and the slope of the treasury curve. In addition they estimate that the negative relationship between credit spreads and the treasury slope is relatively stable over time.

The empirical literature to date supports both the significance and the direction in which structural model variables influence credit spreads, however, recent studies demonstrate that these variables alone are not sufficient to fully explain either the levels or changes in credit spreads. Collin-Dufresne et al. (2001) regress credit spread changes of individual bonds on the changes in treasury yields, the slope of the yield curve, equity index returns, and index volatility, estimating that these variables explain only about 25% of the variation in credit spreads. In addition, the slope of the yield curve is not a significant determinant of credit spread changes when the other variables are taken into account. Using principal components analysis on the residuals they find that the changes in residuals across individual bonds are dominated by a single common systematic component that has no obvious relationship to variables from the interest rate and equity markets. Their conclusion is that yield spread changes are only partly accounted for by the economic determinants of default risk.

7.3 Credit Spread and Default Risk

To estimate how much of the yield spread levels can be accounted for by default risk, Huang and Huang (2003) calibrate a diverse set of structural models to actual historical default losses then use them to generate theoretical values of credit spreads. In each case the model-based spreads are well below the average observed spreads, suggesting that default risk accounts for only a small fraction of yield spreads. The proportion explained by default risk is highest for low-rated bonds, and decreases for higher-rated bonds that have low historical default losses. The inability of theoretical risk variables to account for most of the levels or changes in yield spreads is sometimes referred to as the 'credit spread puzzle'. Similar to the problem of the equity premium puzzle, the expected returns on corporate bonds, like equities, seem well above those justified by the risks.

The explanation of the credit spread premium puzzle has focused on both the presence of additional risks as well as associated risk premiums. Elton, Gruber, et al. (2001) estimate that expected loss accounts for less than 25% of the observed corporate bond spreads, with the remainder due to state taxes and factors commonly associated with the equity premium. Similarly, Delianedis and Geske (2002) attribute credit spreads to taxes, jumps, liquidity and market risk. Factors associated with the equity premium include the Fama and French (1996) 'High-minus-Low' (HML) factor, found by Huang and Kong (2003) to account for a significant component of low-grade credit spread changes. The significance of Fama-French factors is also supported by Joutz et al (2001), who conclude that credit spreads are determined by both default risk and systematic market risk.

Structural models contain the assumption that default risk is diversifiable, since yield spreads are assumed to reflect only default loss, with no risk premium for either default risk or the risk of market-wide changes in spreads. Jarrow et al (2001) show that jumps-to-default in credit spreads cannot be priced if defaults across firms are conditionally independent and if there is an infinite number of firms available for diversification. One explanation for the credit spread puzzle is the potential for firms to default on a wide scale not seen historically, a risk that is difficult to eliminate by diversification and is therefore priced by investors. It is also observed that defaults across firms tend to be correlated and concentrated around recessions when investors are most risk averse.

Chen (2007) and Chen et al (2008) conclude that in order for structural models to capture observed spreads it is necessary to incorporate a strongly cyclical market price of risk that increases along with default losses during recessions. Another explanation for wide credit spreads is that idiosyncratic risk is priced as well as systematic risk. Amato and Remolona (2003) argue that due to the highly skewed returns in corporate bonds, full diversification requires larger portfolios than typically needed for equities. Given the limited supply of bonds, high transaction costs, and possible constraints on portfolio expected losses, full diversification is difficult to achieve, and it is possible that in practice portfolio managers require a risk premium associated with individual bond value-at-risk.

Recent studies confirm the presence of both a firm-specific default risk premium as well as a market risk premium. Driessen (2005) distinguishes between the market-wide changes in credit spreads and individual credit spread jumps-to-default, finding that both components are priced. Similar results are obtained by Gemmill and Keswani (2008), who show that most of the credit spread puzzle can be accounted for by the sum of a systematic risk premium and a larger idiosyncratic risk premium. While supporting these conclusions, Collin-Dufresne et al (2003) also suggest that it is not surprise default itself that attracts a significant premium, but rather it is the potential for credit events of large individual firms to trigger a flight to quality in the treasury market and cause market-wide increases in credit spreads. So even without directly violating the assumption of conditional independence of defaults across firms, idiosyncratic default risk could matter due to its potential to impact market-wide liquidity, which highlights the difficulty of separating the role of default and liquidity in driving credit spread levels.

7.4 Credit Spread and Liquidity

A recent stream of literature focuses on the role of liquidity in explaining both the levels and time-variation of credit spreads. The idea that liquidity is an important and priced determinant of yield spreads is not new, with Fisher's (1959) hypothesis being that the risk premium of individual bonds consists of two main components: a default component and a 'marketability' component. The default component is in part a function of a firm's earnings variability and debt ratio, measures that directly correspond to the leverage and asset volatility variables in structural models, while the marketability component is a function of the outstanding issue size. There is no universal proxy for liquidity risk, but measures used in previous studies include the bid-ask spread, trade frequency, the proportion of zeros in the time-series of a bond's returns, a bond's age, amount outstanding and term to maturity.

Perraudin and Taylor (2003) estimate that liquidity premiums are at least as large as market risk premiums and far larger than expected default losses. De Jong and Driessen (2006) estimate the liquidity risk premium on US corporate bonds at 0.6% for long-maturity investment-grade bonds and 1.5% for speculative-grade bonds. Recent studies estimate the non-default component of credit spreads directly by subtracting credit default swap (CDS) premiums from corresponding corporate bond yield spreads. Being simply contracts, CDS are regarded as more pure reflections of credit risk. CDS are not subject to the same accessibility issues as physical bonds and difficulties in taking a short position. The absence of coupons also avoids bond-specific taxation considerations. This idea of pure credit risk is also supported by the findings of Ericsson et al (2005) that CDS premiums are driven by the theoretical determinants of credit risk (the risk free rate, leverage and volatility), but that in contrast to the results of Collin-Dufresne et al (2001) on corporate bond yield spreads, there is only weak evidence of a common residual factor affecting CDS premiums. The CDS-based non-default component estimated by Longstaff et al (2005) is strongly time-varying and related to both bond-specific and market-wide measures of illiquidity.

There is a strong cross-sectional relationship between the non-default component and individual measures of liquidity such as the bid-ask spread and issue size. The time-series variation of the non-default component is related to macroeconomic or systematic measures of liquidity risk such as i) the spread between on-the-run and off-the-run treasury yields, ii) flows into money market funds, and iii) level of new issuance into the corporate bond market. The cross-sectional results of Longstaff et al (2005) are consistent with equity market evidence of Amihud and Mendelson (1986) that market-average returns are an increasing function of bid-ask spreads, while the time-series results are consistent with the presence of a single common systematic factor found by Collin-Dufresne et al (2001), as well as evidence of systematic liquidity risks in interest rate markets in Duffie and Singleton (1997), Liu et al (2004) and Longstaff (2004). Liquidity risk itself has also been found to be a positive function of the volatility of a firm's assets and its leverage, the same variables that are seen as determinants of credit risk (Ericsson and Renault, 2006).

7.5 Alternative Approach to Analyzing Credit Spread

Our aim in this chapter is to estimate the factors driving the dynamics of yield spread levels directly from the data, without prior assumption about the specific economic variables that yield spreads could be related to. Based on existing evidence, we take the view that the time-variation in credit spreads is driven by two classes of factors that are non-stationary and mean-reverting, respectively. Our initial guess is that the first group of factors is likely to relate to default risk and have low rates of mean-reversion that reflect relatively persistent macroeconomic conditions. The second group could relate to liquidity premiums that are presumed to change with noisy short-term supply and demand shocks. Given that credit risk explains a lower proportion of high-grade spreads than low-grade spreads, we would then expect high-grade spreads to have stronger mean-reversion that reflects changes in liquidity due to supply/demand.

However, from Figure 7.1 it appears that non investment-grade spreads have far more noise than investment-grade spreads, suggesting that the default risk component may be more highly mean-reverting than the remaining component. One indication this may be true for corporate yield spreads is the study of swap spreads by Liu et al (2006), finding time-varying components relating to both liquidity and default risk, but where the default component is highly mean-reverting and with a flat term structure, while the liquidity component is more persistent and with a steep upward-sloping term structure. It is worth noting that spikes in the lowest-grade spread indices resemble the behavior of the VIX index over the same period. These could be interpreted as short-term increases in default risk under the frictionless market framework of structural models, but in practice sharp increases in the VIX are also closely correlated to declines in liquidity. Without assumption about the source of variations, we observe that while the two bond classes behave in fundamentally different ways during particular sub-periods, they also appear to have different exposures to shared common short-term shocks throughout the sample period.

This study assumes that the time-variation in credit spreads across ratings classes is driven by a common set of unobservable factors to which each observed spread is exposed with some unknown sensitivity. We aim to answer the following questions: 1) how many factors are required to explain the evolution of ratings-based spread indices; 2) what is the exposure of each individual index to each factor, and 3) what economic variables, if any, could be proxies for the factors.

Our choice of the state-space methodology is motivated by its advantage of allowing for both time-series and cross-sectional data simultaneously. It also provides a new and opposite approach to the existing literature on credit spread determinants. Most empirical studies on credit spreads adopt a general-to-specific approach where a range of known potential determinants is tested for statistical significance using OLS regressions. In contrast, state-space models require only an assumption about the structure of the factors that can then be estimated directly from the observed data. Another advantage of state-space models is that they can be applied to both stationary and non-stationary variables. OLS estimation on the other hand requires that both dependent and independent variables are stationary, forcing most studies to focus on explaining the changes in credit spreads as a function of changes in independent variables. In this study we analyze the dynamics of credit spread levels directly.

Given an assumed parametric process form for the latent factors, the Kalman Filter maximum likelihood method can be applied to simultaneously estimate 1) the parameters of each factor process, 2) the sensitivities or loadings of each observed series to the individual factors, 3) the realizations of the factor series, and 4) the covariance matrix of the model errors. The Vasicek (1977) mean-reverting process is chosen for the factors since, depending on the size of its mean-reversion coefficient, it is suitable for representing both non-stationary (presumed macroeconomic) as well as stationary (presumed microeconomic) determinants of credit spreads. A multi-factor Vasicek form is also supported by the findings of Pedrosa and Roll (1998) that Gaussian mixtures can capture the fat-tailed distributions of credit spreads.

Early applications of the state-space model in finance literature have focused on the term structure of treasury rates. Babbs and Nowman (1999) find that a three-factor Vasicek model adequately captures variations in the shape of the treasury yield curve, with two factors providing most of the explanatory power. Chen and Scott (1993) and Geyer and Pichler (1999) reach similar conclusions based on a multifactor CIR (1985) model, and find the factors to be closely related to the short rate and the slope of the curve. Recent studies build upon the two or three-factor term structure of treasury rates and allow for additional factors to explain swap or corporate bond yields. Liu, Longstaff, and Mandell (2006) separate the liquidity and credit risk components of swap spreads through a five-factor model of swap yields. Swap yields consist of three factors driving treasury yields, one influencing credit risk, and the remaining one influencing liquidity risk.

Similarly, Feldhutter and Lando (2008) decompose the factors driving the term structure of the swap yield spreads into three factors driving the risk-free rate, two affecting credit risk and one relating to the liquidity premium or 'convenience yield' contained in treasury yields over the risk-free rates. They find that while credit risk is important, the strongest determinant of swap spreads is the convenience yield contained in treasury prices. Jacobs and Li (2008) use the state-space approach to estimate a reduced-form model of default, where the probability of default is modeled directly as a stochastic volatility process. They find that the addition of a second, volatility factor to the level factor in the diffusion of default probabilities leads to significant improvements in both in-sample and out-of-sample fits for credit spreads.

This work is a natural progression in the application of state-space methodology from treasury yield levels to corporate yield spreads. We apply the state space methodology directly to credit spreads to find both the number of factors and compare their behavior to well-known macroeconomic variables.

We emphasize that our choice of the Vasicek factor form is the basis for an econometric representation of credit spreads, rather than a pricing model of spreads across ratings. We directly estimate the factors as well as the exposures of observed spreads to each factor. This is in contrast to the pricing approaches of Babbs and Nowman (1999) and Geyer and Pichler (1999), where yields across maturities are linked by a closed form function of the factor process parameters and maturity.

7.6 Data Used

All data is from Bloomberg with observations taken at the end of each month Apr-96 to Mar-08. We use the 10-year maturity U.S. industrial corporate bond yield indices of 14 available ratings: AAA, AA, A1, A2, A3, BBB1, BBB2, BBB3, BB1, BB2, BB3, B1, B2, and B3. Bloomberg ratings are composites of S&P and Moody's ratings, with bonds rated BB1 or lower considered sub-investment grade.

The yield indices are converted into credit spreads by subtracting the 10-year benchmark bond yield from each. Other variables sourced are the option-implied volatility index of the S&P500 (VIX) and the S&P500 level.

7.7 Multifactor Model for Credit Spread

For a given term to maturity, each of n observed credit spread indices by rating $R_t = \{R_{1t}, R_{2t}, ..., R_{nt}\}'$ is expressed as a function of m independent latent factors or states $X_t = \{X_{1t}, X_{2t}, ..., X_{mt}\}'$ of the Vasicek form. Changes in the j-th observed series R_{jt} are a linear combination of the changes in m latent factors X_{it} weighted by factor loadings $a_{j1}, a_{j2}, ..., a_{jm}$. Each factor evolves according to its three parameters: the long-term mean θ, the speed of mean-reversion κ, and the volatility σ.

The mean-reversion parameter κ is directly related to the time taken for the process to reach its long-run mean θ. In the absence of random shocks the difference between the current level and the mean decays exponentially towards zero. The expected time it takes for the process to decay halfway towards its mean is its "half-life", equal to $\ln(2)/\kappa$ years. In continuous time,

$$dR_{jt} = \sum_{i=1}^{m} a_{ji} dX_{it}, \; j = 1,2,...,n \tag{7.1}$$

$$dX_{it} = \kappa_i \left(\theta_i - X_{it} \right) dt + \sigma_i dW_{it}, \; i = 1,2,...,m . \tag{7.2}$$

The application of the Kalman Filter algorithm to estimate the factor loadings, the process parameters $\psi = \{\kappa_i, \theta_i, \sigma_i\}$ $i = 1,2,...,m$ and the realization of the state vector over time $X = \{X_1, X_2, ..., X_T\}$, requires that the model is expressed in state space form. State space representation consists of the measurement equation and the transition (or state) equation as discussed in Chapter 1.

$$R_t = D + ZX_t + \varepsilon_t, \; \varepsilon_t \sim N(0,H) \tag{7.3}$$

$$X_t = C(\psi) + \Phi(\psi)X_{t-1} + \eta_t, \; \eta_t \sim N(0, Q(\psi)) . \tag{7.4}$$

The measurement equation (7.3) maps the vector of observed credit spreads $R_t(n \times 1)$ to the state vector $X_t(m \times 1)$ via a 'measurement matrix' $Z(n \times m)$ and vector $D(n \times 1)$. Unexpected changes and errors in the sampling of observed series are allowed through n jointly normal error terms $\varepsilon(n \times 1)$ that have zero conditional means and covariance matrix $H(n \times n)$. Since the computational burden of estimating a full error covariance matrix H increases rapidly with additional observed series, most studies assume error independence. In state-space models of the treasury curve, (Chen and Scott (1993), Geyer and Pichler (1996), and Babbs and Nowman (1999)) a diagonal matrix with elements $h_1, h_2, ..., h_n$ was used to capture the effects of differences in bid-ask spreads across n maturities. In this study we choose the same form to allow for different bid-ask spreads across n bond quality groups. The state equation (7.4) represents the discrete-time conditional distribution of the states. The terms of the equation follow directly from the discrete form of the Vasicek model for interval size Δt :

$$X_{i,t+\Delta t} = \theta_i \left(1 - e^{-\kappa_i \Delta t}\right) + e^{-\kappa_i \Delta t} X_{i,t} + \eta_{i,t} \tag{7.5}$$

$$\eta_{i,t} \sim N\left(0, \frac{\sigma_i^2}{2\kappa_i}\left(1 - e^{-2\kappa_i \Delta t}\right)\right). \tag{7.6}$$

Innovations in the states occur through the normal 'noise' vector η_t, with covariance matrix Q. It is assumed that the sources of noise in the state and measurement equations are independent.

In state-space representations of affine models of the term structure, where the observed series correspond to specific maturities, the elements of the measurement matrix Z and the intercept vector D are usually closed-form functions of the term to maturity, the parameters of each risk factor, factor correlations, and the market risk premium associated with each factor.

The difference in this study is that the observed series represent different ratings for a single maturity, without a prescribed formula linking the observed series via factor process parameters. Instead, we estimate the measurement matrix directly by maximum likelihood, along with the process parameters. To reduce the number of parameters in the optimizations we also make the simplifying assumption of a zero intercept vector D. Based on numerous experiments we find no observable impact of this assumption on either the estimated factor realizations or the sensitivities of the observed series.

7.8 Fitting the Model

The state parameters ψ, the elements of the measurement matrix Z, and the measurement error covariance matrix H are estimated by maximizing the log-likelihood function (following directly from Chapter 1), given guesses for ψ, Z, and H, and fixed initialization values X_0, and Σ_0, the log-likelihood is

In maximizing the log-likelihood function we force all the factor loadings of the first observed credit spread series (AAA) to equal 1, so that the first observed series is a non-weighted sum of the latent factors. We add this assumption as a way of ensuring that loadings and factor realizations are scaled comparably across factors and across models with different numbers of factors.

7.9 Results

One, two, and three-factor models are estimated for the period Apr-96 to Mar-03 as well as the full sample period Apr-96 to Mar-08. We are interested in how model estimates are impacted by the changing economic environment. From Apr-96 to Mar-03 lower-grade credit spreads generally increased until reaching their peak in Mar-03 (Figure 7.1). In the period that followed credit spreads generally narrowed and remained low until 2007. The full sample period includes three major shocks to liquidity: the LTCM crisis of 1998, the bursting of the technology bubble and increase in corporate default rates in 2002, and the sub-prime mortgage crisis starting in 2007.

7.9.1 Results for Apr-96 to Mar-03

Table 7.1 shows the estimates for the mean-reversion speed (κ), mean (μ), and volatility (σ) of each Vasicek factor. The log-likelihood, AIC, and BIC criteria are highest for the three-factor model, under which all parameters (with the exception of one mean) are highly significant.

The marginal improvement in the log-likelihood from the addition of a third factor is far smaller than for a second factor, suggesting that a 3-

factor model is sufficient in capturing the common sources of variation in credit spreads.

For comparison, the log-likelihoods for the one, two, three, and four-factor models are 1246.0, 1840.6, 2040.6, and 2100.10, respectively. The parameter estimates for factor 4 in a four-factor model are largely insignificant, supporting the choice of the three-factor model. A parameter is significant at the 5% level if the estimated parameter divided by its standard error is greater than 1.96 in absolute value. The standard errors of the estimated parameters are calculated using a finite-difference estimate of the Hessian matrix, as outlined in Hamilton (1994).

Figure 7.2 shows that the extracted factor under the one-factor model resembles a weighted average of the 14 observed series. Allowing for a second factor reveals two distinct smooth processes as the drivers of the cross-section of credit spreads, while in the three-factor model an additional more noisy process is identified. In the three-factor model the half-life is 2.8 months for factor one, 4.1 years for factor two, and 1.6 years for factor three. The factors under the three-factor model are compared to well-known economic time-series in Figure 7.3.

Under the three-factor model, the noisy first factor resembles the VIX for most of the sample period, the second resembles the (negative of) 10-year bond rate, and the third the S&P500 level. S&P500 returns rather than levels are a more appropriate explanatory variable for credit spreads, since S&P500 levels are strongly upward trending over the long term while yields spreads tend to mean-revert. However, for the sample period used the S&P500 level is roughly stationary, and we compare factors to the levels rather than arbitrary measures of rolling returns. The correlations are 0.08 between factor 1 and the VIX, -0.74 between factor 2 and the long bond rate, and 0.92 between factor 3 and the S&P500 level.

If a "burn-in" phase of the first 12 months is excluded under the Kalman Filter approach, the correlation between factor 1 and the VIX increases from 0.08 to 0.47. Given the results of Campbell and Taksler (2003) linking credit spreads to the average of individual firm volatilities, it is possible that factor 1 is more closely related to measures of the average of individual firm implied volatilities than it is to the VIX which measures the volatility of the market average returns.

The estimated loadings of the observed series to each factor under the one, two and three-factor model are shown in Figure 7.2. Are the sensitivities to the factors consistent with theory? The shape of the loadings on the first factor suggests that equity volatility risk has a positive impact on all credit spreads and that exposure to it increases with declining credit quality. To the extent that equity volatility is a proxy for a firm's asset value volatility, this result is consistent with the prediction of Merton (1974) that the probability of default and credit spreads increase with higher asset value volatility. The sharpest increase occurs in the crossing from investment to sub-investment grade bonds, which is consistent with the observations of Huang and Kong (2003) and others that lower-grade bonds are more sensitive to equity market variables than high-grade bonds.

The positive loadings on factor two and its negative correlation with the level of the 10-year treasury yield are consistent with the strong empirical that increases in treasury yields lower credit spreads. The loadings are also consistent with the finding of Colin-Dufresne et al (2001) that the sensitivity of credit spread changes to interest rates increases monotonically across declining rating groups.

The sensitivities to factor 3, which is closely correlated to the S&P500, change sign from positive to negative as bonds move from investment to sub-investment grade. The estimated positive relationship between equity market performance and investment grade spread indices is at odds with the Merton (1974) model since according to the model, higher equity values increase the value of a firm's assets relative to its fixed level of debt, lowering its probability of default.

A possible explanation is that the positive equity performance throughout the 1990s coincided with rising aggregate debt levels during the same period, with highest rated firms raising their leverage the most. The negative effect of higher asset values on spreads may have been more than offset by the positive effect of higher leverage in the case of higher-grade firms. Changing investor risk preferences may also have played a role. It is possible that for all but the lowest credits, prolonged positive equity market performance contributed more to the substitution out of corporate bonds, in favor of equities, than to higher bond values through improved creditworthiness.

7.9.2 Results for Apr-96 to Mar-08

We repeat the analysis for the full sample period with results reported in Table 7.1 and Figure 7.2. The signs of the correlation coefficients between the factors and macroeconomic variables in the three-factor model remain the same as for the first period: factor 1 and the VIX at 0.71; factor 2 and the long bond rate at -0.54; factor 3 and the S&P500 at 0.76. The general shapes of the loadings and their signs remain unchanged for the 3-factor model, with the exception that loadings on factor 3 are more strongly negative for non-investment grade debt for the full period.

This reflects changing market conditions between the first and the second period. Low-grade credit spreads increased throughout Apr-96 to Mar-03, while the S&P500 reached its peak in mid-2000 and declined until Mar-03. The lack of a strong direction in the relationship between low-grade spreads and the equity market is reflected in the estimated loadings of low-grade spreads on factor 3 being close to zero for the first period. For most of the period that followed (Apr-03 to Mar-08) non-investment grade spreads steadily declined, with lowest grade spreads declining the most, while at the same time the S&P500 trended upwards.

This feature most likely contributes to the estimated loadings of low-grade spreads being more negative and varied across ratings when based on the full sample period. There is also a change in the shape of the loadings on factor 2 for the full period. The loadings peak for the highest-rated non investment grade index (BB1) but then slowly decline with worse ratings. This is in contrast to the finding of Collin-Dufresne et al (2001), supported by our estimates for Apr-96 to Mar-03, that interest rate sensitivities increase monotonically with declining ratings. We note that for the full period factor 2 is less closely correlated to the long bond yield (coefficient of -0.54), than for Apr-96 to Mar-03 (coefficient is -0.74), and that the factor loadings across the two periods are therefore not entirely comparable.

However, the shape of the loadings for the full period raises the question of whether for indices of lower quality than those covered in Collin-Dufresne et al (2001) and this study, the sensitivities to interest rates would decline further across declining ratings. The possibility is also raised by the findings of Fridson and Jonsson (1995), that there is no significant relationship between high-yield spreads and treasury levels.

We find none of the extracted factors in models containing between one to four factors to be correlated to the slope of the treasury curve, either in the spot or forward yields, contemporaneously or with a lag. This is consistent with the findings of Collin-Dufresne et al (2001) that the treasury slope does not help explain credit spread changes, but the result remains surprising given that the treasury curve is commonly used as an indicator of future economic conditions by market participants.

One explanation is that the slope of the treasury curve contains no useful information beyond that already contained in the combination of equity returns, volatility and interest rate levels. Another is simply that the period Apr-96 to Mar-08 contains highly contrasting relationships between credit spreads and the treasury slope, due to the sub-prime crisis. The period since August 2007 has been marked by rapidly widening credit spreads while at the same time fears of stagflation, high inflation and low growth, contributed to short-term treasury yields reaching historically low levels relative to long-term yields. Hence the end of the sample period is marked by a strong positive contemporaneous relationship between the slope of the treasury curve and general credit spread levels, which is in contrast to the negative relationship previously documented by Papageorgiou and Skinner (2006).

We examine the estimated measurement error variances, defined as the diagonals of matrix H in the Measurement Equation (7.3). Figure 7.4 shows the square-roots of the error variances across rating classes for the one, two, and three-factor models. As expected, for each model the lowest-rated bonds which have the widest credit spreads also have the highest estimated measurement error variance. The variances fall sharply across ratings with the addition of a second factor, particularly for sub-investment grade bonds. The addition of a third factor does not lead to a large reduction in variances, which is similar to the modest impact of a third factor on the maximum likelihood.

We note that in the two and three-factor models the measurement error variances also peak around the middle ratings, which is in contrast to the more monotonic shapes of the loadings on each factor, across ratings. The variances increase as indices approach the cross-over point between investment and non-investment grade bonds, with a local maximum for the BB1 index which is the highest-rated non-investment grade index. Both the shape and magnitude of the error variances are comparable to the results of Babbs and Nowman (1999), where a multifactor Vasicek model is used to fit the term structure of 8 observed treasury yields across maturities (0.25, 0.5, 1, 2, 3, 5, 7, and 10-year). In that study, the errors on longer maturities are the highest and decline sharply with the addition of a second factor. The middle maturities around the two-year series have higher error terms than the surrounding maturities.

We provide two possible explanations for the pattern in the error variances. Firstly, the time-series of the first two factors are closely related to averages of the investment-grade bonds and non-investment grade indices, respectively. We would expect that the further an observed index is from the 'average' investment-grade or non-investment grade series, the less precisely it will be captured by the two first and most important factors.

A second explanation is that the observed BB1 index is a relatively noisy proxy for the yield spreads of BB1 quality. From Figure 7.1 it can be observed that the BB1 index closely follows investment-grade bonds early in the sample period, but follows non-investment grade bonds more closely for the remainder of the period. This changeover could be related to changes in the composition of the BB1 index, or changes in the pricing of included bonds that does not get reflected by rating changes. The finite sample of bonds within any rating class creates the potential for measurement errors in the relative pricing of various rating indices, and it is possible that these are more pronounced for indices near the cross-over point between investment and non-investment grade.

The fact that the estimated measurement error variances peak around the crossing point of rating classes while the factor loadings remain relatively smooth can be interpreted as the effectiveness of state-space models in separating the idiosyncratic and systematic effects. Given a wide enough cross-section of time-series, the features of individual series that are not common to multiple indices can be expected to be absorbed into higher measurement error variances, while leaving factor loadings relatively smooth across the series.

7.9.3 Model Performance

We also analyze the time-series of the fit errors under a three-factor model. Table 7.2 shows that for each index the average fit errors are close to zero and strongly stationary based on ADF unit root tests. We take this as support for the multifactor Vasicek as an unbiased model of credit spreads across ratings over the sample period.

In Table 7.3 we report that for at least 12 out of the 14 index time series the hypothesis that fit errors are normally distributed cannot be rejected based on the Chi-squared and Kolmogorov-Smirnov normality tests. We take this as evidence that multiple Vasicek factors provide a realistic representation of the evolution of credit spread indices across ratings. In addition to measuring correlations, we perform a two-sample Kolmogorov-Smirnov test of the hypothesis that each extracted factor and its compared economic variable are independent samples from the same continuous distribution. We note that the changes in each factor are conditionally normal by definition, but the realizations of the factor levels need not be normal.

Table 7.4 shows that the null hypothesis of distributional equivalence cannot be rejected between Factor 1 and the VIX, the (negative of) Factor 2 and the 10-year treasury yield, and Factor 3 and the S&P500.

7.9.4 Discussion

The results for both periods suggest that all credit spreads vary in response to three common systematic factors that have proxies in the VIX, the long bond rate, and S&P500 returns. The co-movement between the factors and the variables is particularly evident from the beginning of the sub-prime crisis. Figure 7.3 shows that from the second half of 2007 factor 1 sharply increased as well as the VIX, factor 2 increased with (the negative of) the long-bond rate, and factor 3 declined with the S&P500 level.

However, the ability of the three factors to explain observed spreads can rapidly decline during financial crises, as shown by the conditional density likelihoods in Figure 7.5. Log-likelihoods dropped during the LTCM liquidity crisis of August 1998, the end of the technology bubble in 2002, and since the start of the 2007 sub-prime mortgage crisis. The implication is that credit spreads reached levels that were not accounted for or fully reflected by the macroeconomic conditions at those times. Interestingly, during the LTCM crisis a two or three-factor model does not improve the fit over a one-factor model.

One interpretation is that this crisis was of a more exogenous nature and more specifically relating to changes in credit market liquidity than changes in the macroeconomic outlook. While the end of the bubble in 2002 and the sub-prime crisis both had long-lasting impacts real economy, reflected in lower yields, lower equity returns and higher volatility, the LTCM crisis was characterized by a relatively sharper increase in volatility and smaller changes in rate and equity returns. It is likely that almost all of the change in credit spread levels during LTCM is explained by the sharp rise in factor 1 which is representative of the VIX, which is in turn closely related to liquidity risk. The sharp falls in log-likelihood that accompany the largest market moves point either to the presence of additional risk factors and risk premiums that are not captured by the Vasicek form, or the need to allow for time-variation in the factor loadings.

7.10 Concluding Remarks

This study concludes that most of the systematic variation in credit spread indices by rating is explained by three factors. The factors vary broadly with the VIX, the long bond rate, and S&P500 returns, which are the theoretical determinants of credit risk. The sensitivities of credit spread indices to each of the factors suggest that the predictions of the Merton (1974) structural model hold on an aggregate level.

While most empirical literature considers liquidity risk, rather than credit risk, to be the major determinant of credit spread levels and changes, we find that the three most important factors driving credit spreads vary with macroeconomic variables. The implication is that the dynamics of a potential liquidity risk premium are not easily separable from those of known macroeconomic variables, a result that is consistent with the findings of Ericsson and Renault (2006) that liquidity risk is determined by the same factors as credit risk. This is the first study to directly estimate credit spread factors using the state-space representation of a multifactor Vasicek model. By making no prior assumptions about the risk variables driving credit spreads, the approach provides a contrast to existing empirical literature and an independent test of theory.

Table 7.1 Parameter Estimates of the One, Two and Three Factor Models.

	April 1996 – March 2003			April 1996 – March 2008		
	1 Factor	2 Factors	3 Factors	1 Factor	2 Factors	3 Factors
Log L	1.246	1.841	2.041	2.316	3.076	3.327
AIC	2.556	3.782	4.217	4.697	6.251	7.790
BIC	2.633	3.903	4.382	4.792	6.400	6.992
Kappa 1	0.391	0.411	2.981	0.120	0.127	0.486
	(0.281)	(0.401)	(0.231)	(0.040)	(0.145)	(0.197)
Kappa 2		0.351	0.171		0.757	0.004
		(5.011)	(0.021)		(0.304)	(0.003)
Kappa 3			0.421			0.786
			(0.041)			(0.357)
Theta 1	0.791	0.601	0.251	1.177	1.195	0.329
	(0.211)	(0.201)	(0.021)	(0.299)	(0.633)	(0.043)
Theta 2		0.161	0.641		-0.010	17.298
		(0.351)	(0.171)		(0.026)	(2.274)
Theta 3			0.161			0.068
			(0.111)			(0.047)
Sigma 1	0.181	0.141	0.121	0.153	0.175	0.105
	(0.021)	(0.011)	(0.011)	(0.011)	(0.011)	(0.009)
Sigma 2		0.141	0.081		0.067	0.126
		(0.021)	(0.011)	(0.022)	(0.011)	(0.011)
Sigma 3			0.131			0.097
			(0.011)			(0.009)

The table shows the maximum-likelihood estimates for each of the three parameters $\{\kappa,\theta,\sigma\}$ of each factor, under the one, two, and three-factor models. The Log-L is the maximized value of the log likelihood function. This is also used to calculate the Akaike Information Criterion (AIC), and Bayesian Information Criterion (BIC). Standard errors based on the inverse Hessian matrix are shown below the parameter estimates.

Table 7.2 Credit Spread Summary Statistics and Model Fit Errors.

Index	Spread Statistics		Error Statistics			ADF Stats. (p-values)	
	Avg.(bp)	SD (bp)	Avg.(bp)	MAPE	SD (bp)	No Trend	Trend
AAA	64.5	25.4	0.01	7.8%	5.7	<0.0001	<0.0001
AA	71.9	27.9	-0.02	5.3%	4.3	<0.0001	<0.0001
A1	83.9	31.3	-0.01	4.3%	4.1	<0.0001	0.0032
A2	94.7	33.7	0.14	2.9%	3.4	<0.0001	<0.0001
A3	108.9	38.4	0.30	3.5%	4.4	<0.0001	<0.0001
BBB1	123.3	40.8	-0.41	5.1%	7.3	0.0020	0.0126
BBB2	138.1	43.2	-0.77	6.2%	9.7	0.0075	0.0350
BBB3	159.9	47.9	-1.36	6.9%	13.0	0.0066	0.0622
BB1	247.1	96.5	4.25	9.5%	24.5	0.0045	0.0221
BB2	292.4	88.5	-0.13	2.5%	8.7	<0.0001	<0.0001
BB3	325.5	98.1	0.54	3.3%	12.8	<0.0001	<0.0001
B1	359.1	99.7	-1.19	2.8%	11.9	<0.0001	<0.0001
B2	415.0	119.6	-0.26	3.7%	19.6	<0.0001	<0.0001
B3	502.8	165.3	4.26	6.6%	39.5	0.0025	0.0112

The table summarizes observed credit spreads for the 144 months from 30-Apr-96 to 31-Mar-08. For each month we calculate a vector of model fit errors, defined as the difference between the 14 observed credit spreads and the fitted spreads defined by the measurement equation in the text. The estimated state vector and measurement matrix are based on the three-factor full period model. For each of the 14 credit spreads by rating we generate a time-series of 144 fit error terms and calculate the average, standard deviation, and mean absolute percentage error (MAPE). Two Augmented Dickey-Fuller tests (ADF) are also performed on each error time-series, one including trend term and the other without. For most of the series the ADF test p-values show that the null hypothesis of a unit root can be strongly rejected at the 5% significance level. The stationary error terms with averages close to zero suggest that the three-factor Vasicek model on average provides an unbiased fit for credit spreads across ratings.

Table 7.3 Normality Test for Model Fit Errors.

| | April 1996-March 2003 | | April 1996-March 2008 | |
	Chi-Sq	KS	Chi-Sq	KS
AAA	0.98	0.97	0.57	0.16
AA	0.49	0.99	0.28	0.26
A1	0.10	0.29	0.40	0.78
A2	0.36	0.45	0.90	0.42
A3	0.65	0.79	0.18	0.26
BBB1	0.69	0.92	0.16	0.31
BBB2	0.04	0.02	0.03	0.19
BBB3	0.24	0.66	0.01	0.09
BB1	0.16	0.49	0.34	0.54
BB2	0.98	0.95	0.96	0.86
BB3	0.87	0.90	0.36	0.23
B1	0.10	0.47	0.99	0.95
B2	0.29	0.65	0.84	0.98
B3	0.31	0.65	0.00	0.01

The table shows the p-values of the Chi-squared and Kolmogorov-Smirnov (KS) tests of the hypothesis that model fit errors are normally distributed. For details about KS test, see Massey (1951). We generate 14 fit-error time series by rating, defined as differences between the observed spreads and model-predicted spreads. We find that for most of the indices the null hypothesis of error normality cannot be rejected at the 5% significance level.

Table 7.4 Kolmogorov-Smirnov Tests of Distributional Equivalence between Factors and Economic Variables.

	Factor 1 Vs. VIX	Factor 2 Vs. -1*10Y Bond	Factor 3 Vs. S&P500
KS Stat	0.097	0.125	0.111
p-Value	0.484	0.196	0.318

We use a three-factor model to generate monthly factor time series for Apr-96 to Mar-08 then linearly transform the factors 1, 2 and 3 to match their means and variances to the VIX, the negative of the 10 year bond yield and the S&P500, respectively. We test the null hypothesis that each transformed factor is drawn from the same distribution as the compared economic variable. The table shows the test-statistics and p-values for the two-sample Kolmogorov-Smirnov (KS) test of the null hypothesis that two independent samples are drawn from the same continuous distribution. For a detailed reference see Chakravarti, Laha and Roy (1967) pp 392 – 394. The test statistic is defined as the maximum difference between the empirical cumulative frequency distributions of two samples. A large value of the test statistic would indicate that two samples are unlikely to be generated by the same underlying distribution. For all three pair-wise tests we find that the null hypothesis of distributional equivalence cannot be rejected at the 5% significance level.

Credit Spread Indices (10-year Maturity)

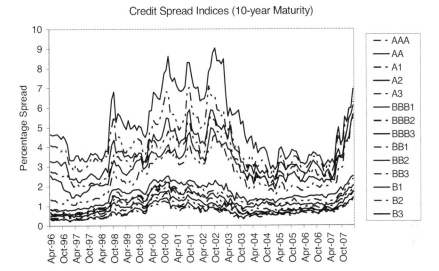

Figure 7.1 US Industrial Corporate Bonds: Credit Spread Indices by Rating.

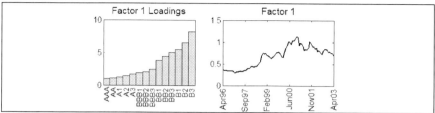

Figure 7.2a Estimated Factor Time Series and the Factor Loadings by Rating:
One Factor Model Period April 1996 – March 2003.

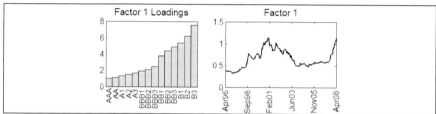

Figure 7.2b Estimated Factor Time Series and the Factor Loadings by Rating:
One Factor Model Period April 1996 – March 2008.

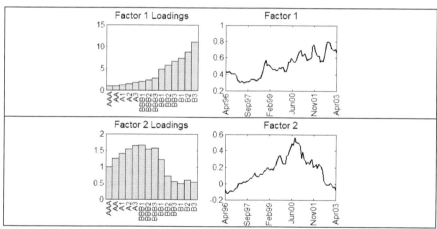

Figure 7.2c Estimated Factor Time Series and the Factor Loadings by Rating:
Two Factor Model Period April 1996 – March 2003.

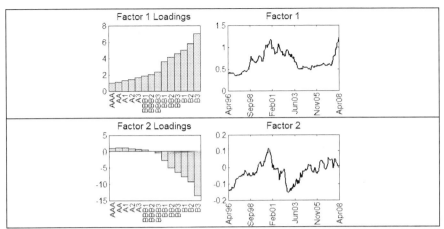

Figure 7.2d Estimated Factor Time Series and the Factor Loadings by Rating:
Two Factor Model Period April 1996 – March 2008.

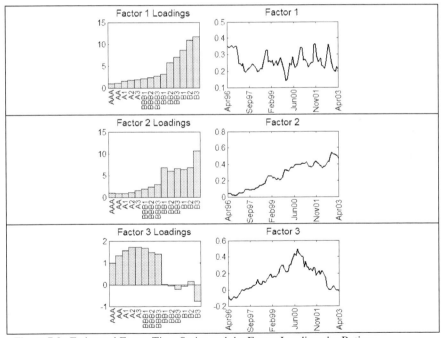

Figure 7.2e Estimated Factor Time Series and the Factor Loadings by Rating:
Three Factor Model Period April 1996 – March 2003.

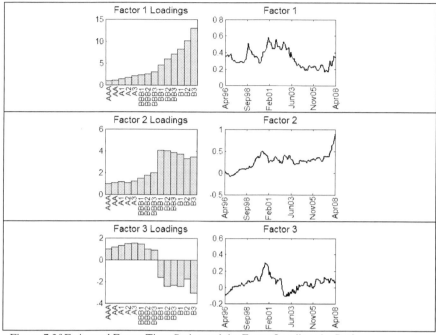

Figure 7.2f Estimated Factor Time Series and the Factor Loadings by Rating:
Three Factor Model Period April 1996 – March 2008.

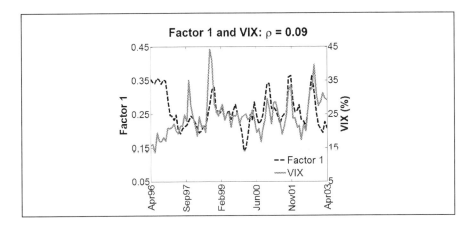

Figure 7.3a Estimated First Factor of the Three-Factor Model and Macroeconomic Variables (Sample April 1996 – March 2003).

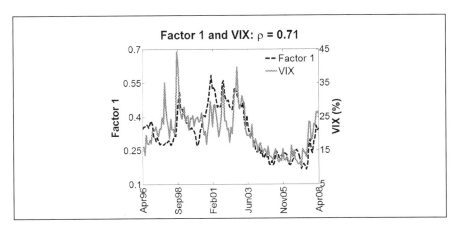

Figure 7.3b Estimated First Factor of the Three-Factor Model and Macroeconomic Variables (Sample April 1996 – March 2008).

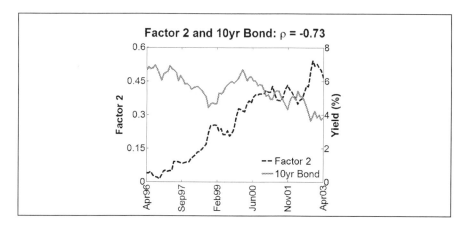

Figure 7.3c Estimated Second Factor of the Three-Factor Model and Macroeconomic Variables (Sample April 1996 – March 2003).

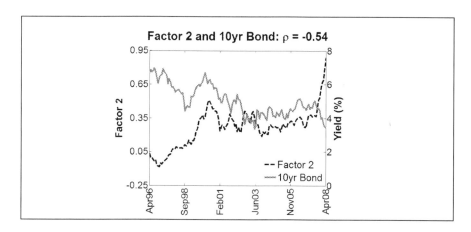

Figure 7.3d Estimated Second Factor of the Three-Factor Model and Macroeconomic Variables (Sample April 1996 – March 2008).

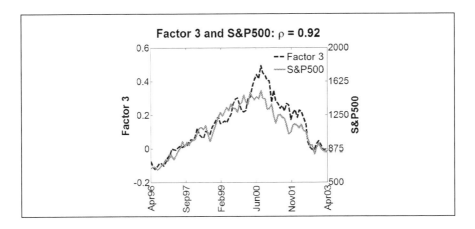

Figure 7.3e Estimated Third Factor of the Three-Factor Model and Macroeconomic Variables (Sample April 1996 – March 2003).

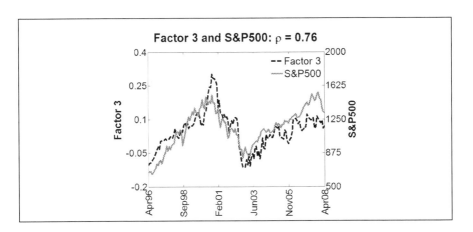

Figure 7.3f Estimated Third Factor of the Three-Factor Model and Macroeconomic Variables (Sample April 1996 – March 2008).

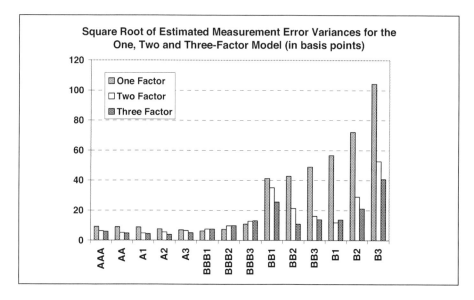

Figure 7.4 Measurement Error Variances (Sample April 1996 – March 2008).

The figure shows the square roots of the estimated measurement error variances for each rating series under the one, two, and three-factor model. All variance estimates are significant at the 5% level within each model.

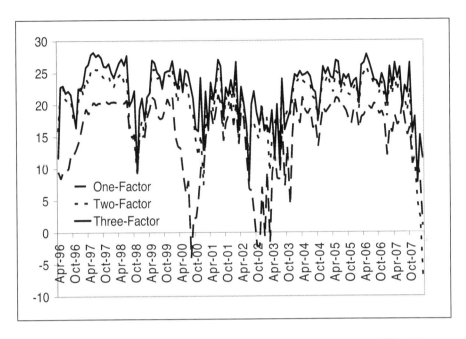

Figure 7.5 Conditional Density Log-Likelihoods for the One, Two, and Three-Factor Model: Apr-96 to Mar-08.

Chapter 8

Credit Default Swaps – Filtering the Components

Credit Default Swaps (CDS) is the most intuitive and highly traded form of the credit derivatives. Its trading grew phenomenally over the last several years, but may have fallen in favor due to recent financial market turmoil. This chapter develops and argues in favor of a component structure of the CDS. These are referred to as long-term and short-term components. These components are extracted by use of stochastic filtering from the observed CDS prices.

The discussion in this chapter also shows why this component structure is necessary to capture various influences from the market in the formation of CDS prices. It goes on to show how these extracted components may be related to other economic variables observable in the market.

The chapter content is based on a paper presented at the 21st Australasian Finance and Banking Conference, Sydney, 16-18 December, 2008, by Bhar, R., Colwell, D. and Wang, P.

We test these components against proxies for several commonly used explanatory variables and find significant but differing impacts of these explanatory variables on the extracted components. For example, equity volatility seems to have a larger influence on the transitory (or short-term) component, suggesting that its effect may be short-lived. At the same time our proxy for illiquidity has a bigger impact on the persistent (or long-term) component, which suggests that its effect is more enduring. Surprisingly, our proxy for the credit rating premium is not even significant in explaining some of the CDS indices, but has a large effect on the persistent component. Finally, the slope of the yield curve has impacts with opposite signs on the two components and thus helps address the conflicting results reported in earlier studies, where such a component framework was not explored. These results indicate that a two factor formulation, similar to Hull and White (1994) interest rate model, may be needed to model CDS options. This will be examined in the next chapter.

Research on credit risk has developed rapidly over the last decade or so. Stimulated by proactive regulatory developments, different classes of models have been put forward to measure, manage, price and forecast credit risk. Coupled with other developments in the credit derivatives market, more incisive research in this market is not only attractive but necessary for better understanding by all participants. Its importance lies not only in risk management activities using Credit Default Swaps (CDS) for hedging purposes, but also for anyone trying to profit from arbitrage possibilities within the CDS market.

8.1 Background to Credit Default Swaps

A credit derivative is an over-the-counter derivative that is designed to transfer credit risk from one party to another. By dynamically adjusting credit exposures, they allow institutions to manage credit risk more effectively. A recent survey conducted by British Banker's Association (BBA) (2006) estimated that by the end of 2006 the size of the market would be $20 trillion and at the end of 2008 the global credit derivatives market could expand to $33 trillion. This growth is expected to continue. Not only is the credit derivatives market growing in total nominal amounts outstanding, it also is attracting an increasingly diversified group of users. Except for the risk management units in large commercial banks, who are the traditional users, investment banks, hedge funds, and other investors are using credit derivatives to profit from arbitrage opportunities in the market. Many are attempting to exploit arbitrage opportunities between the credit derivatives market and the underlying bond and stock market, to hedge positions taken in other markets and also for pure speculation or regulatory arbitrage (Bystrom (2006)).

While there are many variations of credit derivatives, the Credit Default Swap (CDS) is among the most intuitive, and is commonly cited as a basic "building block" for more complex structures. The owner of a CDS receives the bond's face value if the underlying bond defaults. Rather than paying for this default insurance in a lump sum, the holder makes periodic payments until the underlying bond either defaults or matures. If one writes these periodic payments as a percentage of the face value per annum, this quantity is known as the CDS spread (or sometimes, as a CDS price). The CDS spread does behave much like a credit spread. In our data (which essentially consists of averages of CDS spreads, as we discuss later) we observe CDS spreads ranging from 20 basis points to 500 basis points, roughly. Note that there are also bid CDS spreads and ask CDS spreads. It is commonly recognized that the credit default swap spread is a comparatively "pure" measure of credit risk (Ericsson et al. (2004)). Therefore, the study exploring the determinants and the dynamics of the CDS plays a central role when evaluating credit risk related securities and projects.

This chapter is closely related to these works mentioned above, but with a difference. Firstly, we decompose the CDS series of two different maturities into its component structures - a short-term factor and a long-term factor. Secondly, we explore the determinants for each of them separately. To the best of our knowledge this analysis is the first such attempt to look at the determinants of credit default swap spreads form the perspective of the component dynamics.

Here we develop an argument why analyzing the CDS spreads from the viewpoint of component structures is meaningful. It may be argued that the common factors should be relevant for all maturities of CDS spreads. However, the incorporation of information from different explanatory variables affects CDS spreads differently in time and speed. For example, bad or negative news has an almost instant impact on credit spread changes while the impact of good or positive news is much more restrained (Hull et al. (2004)). Also, Norden and Weber (2004) find that the CDS spreads are influenced by the level of the old rating, previous rating events and the pre-event average rating level of agencies. Obviously, the time these influences last is different based on their previous credit rating. For credit default swap spreads with underlying entities having higher credit ratings, the influence of micro and macro elements does not differ much in the short-term and the long-term. On the other hand, for credit default swap spreads with underlying entities having lower credit ratings, the impact of these elements should differ in the short-term and long-term. The credit default swap spreads with underlying firms having lower credit ratings will be more sensitive to the bad news than that of their higher credit rating counterpart firms in the short-term.

Another example of differing influences could be seen as follows. The probability that a firm, which is currently in a rather low credit rating category, would be upgraded is higher in the long run compared to that of a firm which is already in a relatively higher credit rating category. Also just as illiquidity could affect bonds with different maturities to a different extent, the illiquidity could also affect long-term CDSs and short-term CDSs to a different extent. Moreover the assessment of overall economic health of investors in the future would influence the long-term factor differently from that of the short-term factor.

Now that we have established the case for a component structure approach, the question is how we infer these components given that we only have observations on CDS spreads. In this respect, we follow the framework of Schwartz and Smith (2000). We include two unobserved components in the model in such a way that one would capture the long-term dynamics and the other one would capture the short-term dynamics.

The long-term factor is thought to describe the equilibrium spread level (representing a constant fundamental spread level) and is assumed to evolve according to a Brownian motion with drift reflecting the business cycle, development of the credit derivatives market, as well as political and regulatory effects. The short-term deviations, which are defined as the difference between spot and equilibrium prices, are expected to follow a mean reverting process. These deviations may reflect, for example, portfolio adjustments or intermittent market disruptions, and temporary credit changes. Since neither of these factors is directly observable, we can set up the problem in an unobserved component framework and use standard Kalman filtering techniques to make optimum inferences about the state variables.

In this chapter we focus on different indices of CDS spreads, namely the iTraxx family which can be treated as portfolios, so the idiosyncratic differences are rendered trivial. We, thus, care more about the common factors behind all these CDS spreads. Also we choose three benchmark iTraxx indices, namely iTraxx Europe, iTraxx Europe Hivol and iTraxx Europe Crossover. This allows us to have a complete view of the European CDS market just by examining these three indices.

Overall, our results show that with the quick development of the CDS market (or credit derivatives market) the level of CDS spreads is falling during the sample period. This may mainly result from the enhanced efficiency and liquidity of this market. For firms in different credit rating categories, the short-term and long-term behavior is different. The credit default swap spread for underlying entities with higher credit ratings tends to be more volatile in the long-term while the credit default swap spread for underlying entities with lower credit rating tends to be more volatile in the short-term. Our results also demonstrate that most theoretical explanatory variables, as documented by other researchers, have significant impacts on credit default swap spreads and they do impact short-term and long-term dynamic of CDS spreads differently. Moreover, these variables can explain more variations of long-term dynamic in CDS spreads compared to those of short-term dynamic.

8.2 What is in the Literature Already?

In the contemporary literature on credit derivatives, theoretical and empirical work has mostly concentrated on estimation of default probabilities and determinants of credit risk. Most of these studies, however, use the corporate bond spread, also called the credit spread. The credit spread is generally defined as the yield of corporate bond minus the yield on a government bond with comparable maturity.

Longstaff and Schwartz (1995a, b) use annual data from 1977 to 1992 in a regression based analysis of the change in credit spread against the change in the 30-year-Treasury rate and the return on an appropriate equity index. Irrespective of maturity, they find the intercept term and coefficients of change of the 30-year-Treasury rate increase in absolute magnitude as the credit quality decreases. Wilson (1997a, b) examined the effects of macro-economic variables, namely, GDP growth rate, unemployment rate, long-term interest rate, foreign exchange rate and aggregate savings rate, in estimating the default rate level. Collin-Dufresne, Goldstein and Martin (2001), on the other hand, investigated the determinants of credit spread changes using dealer's quotes and transition prices on industrial bonds. They show that the proxies for three theoretical determinants namely, default risk, recovery rate upon default and liquidity have rather limited explanatory power. Their results actually suggest that monthly credit spread changes are mainly driven by local supply/demand shocks that are independent of both credit-risk and standard proxies for liquidity.

Studies which directly focus on the determinants of CDS spreads are also developing rapidly in recent years. Skinner et al. (2003) use arguments from option pricing theory and suggest that the CDS spread should be highly dependent on the risk-free short rate, the yield of the reference obligation, the interest rate volatility, the time to maturity and the recovery rate. They find that four of these variables, namely the risk-free short rate, the yield of the reference obligation, the interest rate volatility and the time to maturity, contain significant information. Benkert (2004) conducts a regression analysis using CDS panel data, incorporating variables such as credit rating, liquidity, leverage, historical volatility and implied volatility. The author finds that implied volatility has a stronger effect than the historical volatility, and that both remain relevant in the presence of credit ratings which contribute an equal amount of explanatory power.

The aim of this chapter is to develop a model that is realistic but simple enough to capture the dynamic behavior of CDS spreads and explore its potential determinants through its component representation. Schwartz and Smith (2000) devise such a component modeling approach by considering short-term variation and long-term dynamics in commodity prices. In their model the natural logarithm of the commodity price is assumed to be the sum of these two components. The short-term component is assumed to revert back to zero based upon its mean reversion speed and the long-term component is assumed to be a random walk with a deterministic trend. In essence, the interpretation of long-term and short-term factors in their model is in line with persistent and temporary components, respectively. Zhou and Qing (2000) take a similar approach, working with logarithmic stock prices.

8.3 Credit Derivatives Market and iTraxx Indices

The structured credit derivatives market encompasses a wide range of capital market products designed to transfer credit risk among investors through over-the-counter transactions. The findings of a recent survey by the British Banker's Association (BBA) have already been described in an earlier section. Figure 8.1 plots the global market size of credit derivatives from 1996 to 2008 (estimated).

This growth may have been driven by several factors. For example, it is well known that credit derivatives are designed to meet the needs of risk management as well as develop credit arbitrage techniques to create new investment products. Under increasing pressure to improve financial performance, banks have turned to the credit derivatives market to diversify their credit risk exposures and to free up capital from regulatory constraints (Duffee and Zhou (2001)). Another reason would be the presence of significant credit arbitrage opportunities across different market sectors, for example, loans and bonds and across different countries. Structured credit entails applying financial engineering techniques to leverage these opportunities and create customized financial products for investors including credit linked notes and repackaged notes.

As mentioned in the section 8.1, while there are many variations of credit derivatives, Credit Default Swap (CDS) spreads are among the most intuitive. From Table 8.1, obtained from the British Banker's Association 2006 Credit Derivatives Report, single-name credit default swaps still represent a substantial section of the market although its share has fallen to 33%. During the same time period, the share of full index trades has increased to 30% as of the first quarter of 2006 and has become the second largest section of the credit derivatives market.

When it comes to the participants in the credit derivatives market, as predicted in the previous survey, hedge funds have become a major force, their share of volume in both buying and selling credit protection having almost doubled since 2004. However, banks still constitute the majority of market participants.

In a manner similar to the way a stock index is created, a CDS index, for example the iTraxx index is a portfolio of single-name credit default swaps which should diversify any diversifiable risk. The iTraxx index family consists of various indices of the most liquid CDS contracts in Europe and Asia (in the US, since April 2004, a similar family of indices is called the Dow Jones CDX). The benchmark iTraxx Europe index comprises 125 equally-weighted investment grade European names which are selected by a dealer poll based on CDS volume traded over the previous six months. Other than the benchmark iTraxx Europe index, there are two other popular benchmark iTraxx indices. These are the benchmark iTraxx Europe Crossover index comprising the 50 sub-investment grade entities from iTraxx Europe and the benchmark iTraxx Europe Hivol index consisting of the 30 names with the widest CDS spreads from iTraxx Europe.

The iTraxx indices typically trade 5 and 10-year maturities and a new series is determined by a dealer liquidity poll every 6 months. The Europe and Hivol indices also trade 3 and 7-year maturities. The indices are managed and administrated by the International Index Company that is owned by a group of the largest global investment banks. All index quotes for our study have been made available by the International Index Company.

8.4 CDS Index Data and Preliminary Analysis

This paper examines the natural logarithm of the CDS index represented by three benchmark iTraxx indices. The theoretical explanatory variables considered in this paper are: the short interest rate, the slope of yield curve, the stock market volatility, the bid-ask spread for each iTraxx index and each maturity and credit rating premia. The data consists of daily time series (749 observations), covering the period June 21, 2004 to May 29, 2007. The iTraxx indices and the level of the interest rates are expressed in basis points, annually. All computations are carried out with the natural logarithm of these variables to be consistent.

The three benchmark iTraxx indices used in this study are iTraxx Europe, iTraxx Europe Hivol and iTraxx Europe Crossover. These three benchmark iTraxx indices give us a complete overview of the whole European CDS market. All these indices are traded with 5-year as well as 10-year maturity and are denominated in Euro.

iTraxx Europe is made up of 125 equally weighted investment grade European names, and is used as a benchmark index. These 125 component names are updated every 6 months by a dealer poll based on CDS volume traded over the previous 6 months. Table 8.4 lists the component names for each of three benchmark iTraxx indices in series 7 which is an iTraxx index time series covering the period from September 20, 2006 to March 19, 2007.

Figure 8.2 plots these iTraxx indices along the time axes and gives readers a rough idea of the dynamic behavior of the CDS index. Table 8.2 gives descriptive statistics for both (the natural logarithm of) the iTraxx index level and its first difference. For example, in logarithmic form, index levels range from 3.0143 to 6.2222, which correspond to CDS spreads of $\exp(3.0143) = 20.4$ basis points to $\exp(6.2222) = 503.8$ basis points.

There are several points we would like to emphasize based on Figure 8.2. Firstly, it is not surprising that the iTraxx index level is monotonically increasing with the maturity which is consistent with the upward slope of the term structure of CDS spreads. Secondly, we notice that there is a less obvious downward trend in our sample period which is quite different from recent credit derivatives markets which suffered from the collapse of the U.S. subprime market. Also, during the sample period, the most obvious downward trend, which is a narrowing of the spread, demonstrated in the first nine months before the problems faced by General Motors (GM) and Ford. Ford and General Motors are two of the world's biggest car companies, as recorded by the web site of www.financialpolicy.org. GM had $290 billion in outstanding debt and Ford had another $160 for a total of $450 billion.

Given the massive size of the auto makers' debt, this turbulence seems to have spread across the Atlantic to the European auto sector and thus affected the whole iTraxx. For example, the iTraxx Autos 5Y was up 16.45 basis points from 65.66 basis points on 5 May, 2005 to 82.11 basis points on 18 May, 2005; while iTraxx Europe 5Y raised from 42.56 basis points to 57.89 basis points during the same period. Despite this turbulence in the credit derivatives market, the overall downward trend in our iTraxx indices during our sample period are observable in our estimation results and a more detailed discussion is given in a later section.

Thirdly, it is interesting to notice that the spreads between the 5-year maturity and the 10-year maturity gets wider with passing time, which may be because the 5-year contract became more actively traded than the 10-year maturity with the development of the CDS market. It also may be the impact of the leverage ratio. The mean-reversion feature (Collin-Dufresne and Goldstein (2001)) in the leverage ratio can significantly increase the credit spreads of long-term debt but has little impact on the short-term credit spreads since the change in the default boundary in the short term is negligible.

Fourthly, turning to Table 8.2, to the extent that the spread is a compensation for credit risk, it is not surprising that the sub-investment-grade firms are considered riskier by the market. Accordingly, the iTraxx Europe Crossover index gives the widest CDS index spread in the sample period which is consistent with theoretical and empirical results that the credit rating is the main determinant of the credit default swap spread. For example, the average credit rating for iTraxx Europe is A2/A2, while it is Ba3/B1 for iTraxx Europe Crossover. More detailed description about credit rating can be found in a later section. The standard deviations for each spread are quite close to each other, with a standard deviation about 219% on an annual basis and this deviation is much higher than that of a stock index.

When it comes to the distribution of daily log differences on iTraxx index and stock index we find that the distribution for a CDS index is much more skewed and leptokurtic than that of log differences on a stock index. This log difference on the iTraxx index is also at least two to three times more volatile compared to stock index return. The volatilities for CDS spreads range from 23.8% - 42.7% on an annual basis, while in the case of stock index the volatility is around 12.6% on an annual basis. All these observations tend to indicate that the CDS market is reacting relatively more strongly to credit deteriorations than credit improvement in comparison to the stock market. With reference to the three iTraxx indices, the iTraxx Europe Crossover index has especially significant large positive skewness and kurtosis. This may indicate that the CDS's with lower previous grade react relatively more strongly to credit deteriorations than credit improvements as suggested by Norden and Weber (2004).

While adopting the framework of Schwartz and Smith (2000) to decompose the iTraxx index into the two components, we need to confirm the non-stationarity of the original iTraxx indices. In order to do that, we perform both the Augmented Dickey-Fuller test and the Phillips-Perron test for these series. For robustness, we try different model specifications and different lags. The results are quite similar and for simplicity, here we just list the results for the unit root test without trend and with four lags. The Unit Root test shows that all iTraxx indices are non-stationary at the 5% significance level while all daily log-difference on iTraxx indices are stationary.

8.5 Focusing on Explanatory Variables

Published literature, for example Ericsson et al (2009) and Abid et al (2006a, b), has argued about the importance of several explanatory variables to determine the market wide movements of credit spreads. In the context of CDS spreads it is customary to use these variables to explain their variations. In this study we also use those variables to analyze the components of CDS spreads. Specifically, the explanatory variables in this analysis include credit ratings, the short-term interest rate, the slope of the yield curve, equity volatility and some measure of liquidity.

Credit Rating:

The premium related to credit rating is a very important determinant of credit related securities including CDS's. Here, we use the difference between the iTraxx Europe Crossover index level and the iTraxx Europe index level to proxy the credit rating premia effect. We would expect it be more related to the short-term factor. As pointed out by Norden and Weber (2004), the abnormal performance in the CDS market is influenced by the level of the old rating, previous rating events and also the pre-event average rating level of all agencies. This means that the credit default swap spreads of the underlying entity with lower credit rating are more sensitive in the short-term. But of course it should have an impact on the long-term factor as well but the extent of this influence should not be as strong.

Credit ratings for sovereign and corporate bond issues have been produced by rating agencies such as Moody's and Standard and Poor's (S&P) for many years. In the case of Moody's the best rating is Aaa. Bonds with this rating are considered to have almost no chance of defaulting in the near future. The next best rating is Aa. After that comes A, Baa, Ba, B and Caa. The S&P ratings corresponding to Moody's Aaa, Aa, A, Baa, Ba, B, and Caa are AAA, AA, A, BBB, BB, B, and CCC, respectively. To create finer rating categories Moody's divides its Aa category into Aa1, Aa2, and Aa3 and so on. Similarly S&P divides its AA category into AA+, AA, and AA-. Only the Moody's Aaa and S&P

AAA categories are not subdivided. Ratings above or on Baa3 (Moody's) and BBB- (S&P) are referred to as investment grade.

Since there is no credit rating assigned to the iTraxx index directly, the rating of iTraxx index in this paper is calculated as the average of the ratings of its component companies. Although each new series of iTraxx index is determined every 6 months and the entity name incorporated in each index may change over different series, the change in average credit ratings is thought to be trivial. So in this study we calculate the average credit rating based on iTraxx Europe Series 7 membership list which can be found on the web site of International Index Company. The rating assessed for each issuer by Moody's and S&P are quite similar and we adopt the rating of Moody's if there is a rating provided; otherwise, we adopt the rating of S&P. We use a numerical equivalent of the credit rating exhibited on Table 8.3. In Table 8.4 we provide the average credit rating for each of our three benchmark iTraxx indices. It is not surprising that the iTraxx Europe Crossover has the lowest credit rating with average Ba3/B1, while the iTraxx Europe has the highest credit rating corresponding to the lowest CDS spread.

Figure 8.3 plots the credit rating premia for both 5-year maturity indices and 10-year maturity indices, which is calculated as the logarithm of the level of the iTraxx Europe Crossover index minus the logarithm of the level of the corresponding iTraxx Europe index. The two curves are almost parallel, which on one hand confirms that the credit rating is a key determinant of a credit default swap spread as well as time to maturity. It also shows that the proxy for the credit ratings formed in this study is quite reasonable.

Short-Term Interest Rate:

It is widely accepted that credit risk cannot be priced independently from market risk, especially interest rate risk (Jarrow and Yildirim (2002)). Duffie and singleton (1999) assume that the intensity of default in reduced form models is a stochastic process that derives its randomness from a set of variables such as the short term interest rate. Several choices can be made for the short interest rate proxy. Ait-Sahalia (1996a, b) used seven-day Eurodollar rates and Stanton (1997) used three-month Treasury bill rates, for example. Taking consideration of market region and data accessibility, we use the 3-month Euribor rate as the proxy for the short term interest rate. We obtained this data from DataStream. Daily observations are used and we apply the 3-month Euribor rate for the day prior to the credit default swap quote. Figure 8.4 plots these 3-month Euribor rates covering our sample period.

A negative relationship between the level of the short term interest rate and the credit spread has been documented for several datasets; see for example Longstaff and Schwartz (1995 a, b) or Duffee (1998). Similarly, Abid and Naifar (2005) find the use of the risk-free interest rate as an explanatory variable increases the total adjusted R-square and the variable risk-free interest rate is negatively correlated to the levels of credit default swap spreads. That is, an increase in the short term interest rate leads to a reduction in the spreads. Generally credit risk results from the default of punctual commitment represented by the principal and interest payments owed to a debt holder. Therefore when we examine its influence on both the short-term and the long term factors, it is reasonable to hypothesize that the short term interest rate will have more influence on the short-term factor as opposed to the long-term factor.

Slope of the Yield Curve:

The slope of yield curve, which can be interpreted as an indication of overall economic health, has the following predicted impact on CDS spreads. Theoretically, an increase in the slope of the yield curve should increase the expected future short term interest rate and as a result, a decrease in CDS spreads if the negative relationship between credit risk and the short term interest rate hold. The slope of the yield curve is measured as the difference between the long term interest rate and short term interest rate. Here we use the European 10-year government bond yield minus the European 2-year government bond yield. All these daily data are download form DataStream and as with the short term interest rate, we take the data for the day prior to the credit default swap quote and plot it in Figure 8.4.

This figure demonstrates two different patterns with the short term interest rate and the slope of yield curve. For the short term interest rate, it is more or less flat during the first 16 months of our sample period and then keeps increasing during the rest of the period while the slope of yield curve goes the other way, decreasing with time passing. When we take a detailed look at the slope of yield curve, we find that the decrease in the slope mainly results from the increasing short-term interest rate, which is the European 2-year government bond yield and the relatively flat long-term interest rate, which is European 10-year government bond yield, during the sample period.

  Conflicting empirical results about the relationship between the slope of the term structure and changes in credit spreads have been documented in previous literature. For example, Collin-Dufresne, Goldstein and Martin (2001) conclude that the slope, calculated as the 10-year interest rate minus the 2-year interest rate is not a significant explanatory variable, either statistically or economically. This is in line with the results by Ericsson et al. (2009) based on credit default swap spreads. However, Abid and Naifar (2005) find a significance negative relationship between credit default swap spreads and the slope of the yield curve. Meanwhile, Avramov et al. (2007) find the slope calculated as 30-year interest rate minus 2-year interest rate to be significantly positively related to changes in credit spreads. Similarly, the slope calculated as 5-year interest rate minus 2-year interest rate is significantly negatively related with changes in credit spreads based on constant maturity yield curves.

  Since the aforementioned studies do not use a component structure, these conflicting empirical results possibly come from the mixed influence of the temporary and persistent components. We argue that our approach in this paper based on component structures may reconcile some of these conflicting observations. So in our later analysis, we would expect to see a different influence of the slope of yield curve on the temporary component and the persistent component.

Equity Volatility:

  The volatility of equity is the main driver of credit risk in structural models. A recent paper by Abid and Naifar (2006a, b) find that the equity return volatility of reference entities can be a proxy for default risk. The volatility they used is estimated from a GARCH (1, 1) model. It is now commonplace to measure volatility in financial time series using GARCH models. These models are based on the notion that the innovations of a time series unconditionally have a fixed variance, but that volatility clustering occurs in the sense that the conditional variance of the process varies over time. The GARCH (1, 1) can be expressed as following:

$$h_t = \alpha_0 + \alpha_1 \varepsilon_{t-1}^2 + \alpha_3 h_{t-1} \qquad (8.1)$$

where ε_t is the innovation in the levels and h_t is the conditional variance on date t .

In this study we adopt the Morgan Stanley Capital International (MSCI) Europe Index (obtained from DataStream) as the representation of the aggregate European equity market. The MSCI Europe Index is a free float-adjusted market capitalization index that is designed to measure developed market equity performance in Europe. As of August 2007, the MSCI Europe Index consisted of the following 16 developed country market indices: Austria, Belgium, Denmark, Finland, France, Germany, Greece, Ireland, Italy, the Netherlands, Norway, Portugal, Spain, Sweden, Switzerland, and the United Kingdom.

In order to estimate time-varying volatility from a GARCH (1, 1) model, we also need to choose a mean equation. In this analysis, we adopt the following return specification in order to generate time varying volatility estimates of the equity market as a whole:

$$r_t = c + \varepsilon_t . \qquad (8.2)$$

This constant c is the mean of the series and ε_t is the residual or the difference between the realized value and the mean. If there is autocorrelation and partial autocorrelation in these squared residual series ε_t^2, it is a signal that the variance is a predictable process. The statistic to be used here to test for autocorrelation and partial autocorrelation is the Ljung-Box Q-Statistic. Our Q-Statistic value is quite large which suggests the existence of autocorrelation and partial autocorrelation. When we regress ε_t^2 on its k lagged values, the results suggest the presence of an ARCH effect. Thus, it is very reasonable for us to use a GARCH (1, 1) model to estimate time-varying volatility.

In order to obtain robust results we test several other time series dynamics (ARMA) for the mean of the equity return data as well as different lag specifications for the GARCH variance part. Using the Akaike (AIC) and the Bayesian (BIC) information criteria we compare the alternative models. We finally choose a model that is as simple as possible given their comparable performance and focus on the constant mean and GARCH (1, 1) version. Intuitively, we expect this equity market volatility to be more related to the temporary component. Figure 8.5 plots the estimated time varying volatility over our sample period.

Bid-Ask Spread (Proxy for Liquidity)

We use the bid-ask spread as a proxy variable for the liquidity factor. It has been widely suggested that financial securities subject to default risk also contain a premia for bearing illiquidity risk; see for example Driessen (2005) for liquidity factors in credit spreads, or Longstaff et al. (2005) for liquidity factors in credit default swaps. Theoretically assets with more liquidity have lower credit spread. A higher bid-ask spread means low liquidity which would leads to a higher CDS spread.

For the liquidity proxy used in this study we calculate the ask quote minus the bid quote and then take its natural logarithm for each iTraxx index with 5-year maturity as well as 10-year maturity. We plot these liquidity proxies in Figure 8.6. Usually, bid-ask spreads of 10-year maturity contracts are higher than those of 5-year maturity contracts for the same iTraxx series. Among all these iTraxx indices, iTraxx Europe Crossover has highest bid-ask spread because of its sub-investment grade components. In our analysis, we would expect to find that the liquidity proxy is positively correlated with the iTraxx index level, especially with the transitory component in our analysis.

8.6 Methodology for Component Structure

8.6.1 Latent-Component Model for iTraxx Indices

Our previous Unit root test shows that all iTraxx indices of interest are non-stationary at the 5% significant level. Similar to Schwartz and Smith (2000) and Zhou and Qing (2000), we describe our natural logarithm iTraxx index as a linear combination of a temporary component and a persistent component. The temporary component is assumed to be stationary and the persistent component is represented by a non-stationary dynamic. We present our model for the dynamic behavior of the natural logarithm of CDS spreads for two different maturities as follows:

$$\begin{bmatrix} CDS_t^5 \\ CDS_t^{10} \end{bmatrix} = \begin{bmatrix} a \\ b \end{bmatrix} + \begin{bmatrix} 1 & 1 \\ 1 & d \end{bmatrix}\begin{bmatrix} X_{1,t} \\ X_{2,t} \end{bmatrix} + \begin{bmatrix} \varepsilon_{1,t} \\ \varepsilon_{2,t} \end{bmatrix}, \tag{8.3}$$

$$dX_{1,t} = -kX_{1,t}dt + \sigma_1 dW_{1,t}, \tag{8.4}$$

$$dX_{2,t} = \mu dt + \sigma_2 dW_{2,t}. \tag{8.5}$$

In the above formulation, CDS_t^T is the natural logarithm of the T period spread at time t with constant maturity T ; X_1 is the temporary and X_2 is the persistent component of CDS_t^T respectively. d is the parameter of persistent component in equation (8.3). $\varepsilon_{i,t}$ is the error term, which is assumed to have an independent identical distribution with mean 0 and variance $\sigma_{\varepsilon,i}^2, i = 1,2$.

The temporary component X_1 follows an Ornstein-Uhlenbeck process which is also known as a mean-reverting stochastic process. It is generally given by the following stochastic differential equation, where θ is the level to which X_1 reverts:

$$dX_{1,t} = k(\theta - X_{1,t})dt + \sigma_1 dW_{1,t}. \tag{8.6}$$

Setting θ in our case equal to 0 and integrating equation (8.6) within interval $[0,t]$ we get,

$$X_{1,t} = X_{1,0}e^{-kt} + \sigma_1 \int_0^t e^{-k(t-s)}dW_{1,s}. \tag{8.7}$$

As a consequence, the temporary component X_1 is normally distributed with mean and variance given by,

$$E[X_{1,t}] = X_{1,0}e^{-kt},$$ (8.8)

$$Var[X_{1,t}] = \frac{\sigma_1^2}{2k}[1 - e^{-2kt}].$$ (8.9)

Notice, indeed, that the drift of the process X_1 is positive whenever X_1 is below θ (which equals 0 here) and negative otherwise, so that at any time X_1 tends to be pushed towards the level $\theta(=0)$. Also, $\theta(=0)$ can be regarded as a long-term average level of X_1. Here, k and σ_1 are constants and are positive. Note that this approach is related to the Hull and White (1994) two-factor model, although one of the factors has no mean reversion in our case.

The persistent component X_2 is assumed to follow a Brownian motion with deterministic drift. For ease of estimation, we assume the noise term $\varepsilon_{i,t}$ is independent of dW_1 and dW_2, but dW_1 and dW_2 are correlated with correlation ρ; that is $E[dW_1 \cdot dW_2] = \rho dt$.

The state variable X_1 may be related to transitory market factors such as noise and feedback trading as that happens in stock market. As a result, changes in the temporary component represent short-term changes in prices that are not expected to persist while changes in the persistent component represent fundamental, long-term changes that are expected to persist.

We rewrite our continuous-time model in discrete form based on the measurement interval Δt. So for equation (8.4) we get,

$$\Delta X_{1,t} = -kX_{1,t-1}\Delta t + \sigma_1 \Delta Z_1,$$

$$X_{1,t} - X_{1,t-1} = -kX_{1,t-1}\Delta t + \sigma_1 \Delta Z_1,$$

$$X_{1,t} = (1 - k\Delta t)X_{1,t-1} + \sigma_1 \Delta Z_1.$$

Similarly we can transform equation (8.5) and get,

$$\Delta X_2 = \mu \Delta t + \sigma_2 \Delta Z_2,$$

$$X_{2,t} - X_{2,t-1} = \mu \Delta t + \sigma_2 \Delta Z_2,$$

$$X_{2,t} = X_{2,t-1} + \mu \Delta t + \sigma_2 \Delta Z_2.$$

If we express it in matrix form, then we get

$$\begin{bmatrix} X_{1,t} \\ X_{2,t} \end{bmatrix} = \begin{bmatrix} 0 \\ \mu\Delta t \end{bmatrix} + \begin{bmatrix} 1-k\Delta t & 0 \\ 0 & 1 \end{bmatrix} \times \begin{bmatrix} X_{1,t-1} \\ X_{2,t-1} \end{bmatrix} + \begin{bmatrix} v_{1,t} \\ v_{2,t} \end{bmatrix}, \qquad (8.10)$$

$$\begin{bmatrix} v_{1,t} \\ v_{2,t} \end{bmatrix} \sim N\left(\begin{bmatrix} 0 \\ 0 \end{bmatrix}, \Omega \right), \Omega = \begin{bmatrix} \sigma_1^2 & \rho\sigma_1\sigma_2 \\ \rho\sigma_1\sigma_2 & \sigma_2^2 \end{bmatrix} \Delta t . \qquad (8.11)$$

The parameter set is. $\Theta \equiv \{k, \sigma_1, \sigma_2, \mu, \rho, a, b, d, \sigma_{1,\varepsilon}, \sigma_{2,\varepsilon}\}$.

The measurement equation (8.3) and the state dynamics (8.10) may be cast into State Space framework, and we get:

$$Y_t = H \times S_t + W_t, W_t \sim N(0,R) , \qquad (8.12)$$

$$S_t = J + F \times S_{t-1} + V_t, V_t \sim N(0,Q) . \qquad (8.13)$$

where,

$$S_t = \begin{bmatrix} X_{1,t} \\ X_{2,t} \end{bmatrix}, J = \begin{bmatrix} 0 \\ \mu\Delta t \end{bmatrix}, F = \begin{bmatrix} 1-k\Delta t & 0 \\ 0 & 1 \end{bmatrix}, H = \begin{bmatrix} 1 & 1 \\ 1 & d \end{bmatrix}, Y_t = \begin{bmatrix} CDS_t^5 \\ CDS_t^{10} \end{bmatrix};$$

$$R = \begin{bmatrix} \sigma_{1,\varepsilon}^2 & 0 \\ 0 & \sigma_{2,\varepsilon}^2 \end{bmatrix}, \text{ and } Q = \Omega \text{ as defined in equation (8.11)}.$$

8.6.2 State Space Model and Stochastic Filtering

State-space framework typically deals with dynamic time series models that involve unobserved components, and are used widely in applied econometric and financial applications. Compared with the multivariate regression models, the state-space model does not require an a priori specification of the predictive variables (Zhou and Qing (2000)). In the literature, considerable effort has been devoted to empirically model the underlying factors describing the dynamic behavior of asset prices.

The basic estimation tool used to deal with the standard state-space model is Kalman filter. The Kalman filter is a recursive procedure for computing optimal estimates of unobserved state variables at time based on available information at time t . Given a prior distribution on the initial value of the state variables and a model describing the likelihood of the observations as a function of the true value, the Kalman filter generates updated posterior distributions for these state variables in accordance with Bayes' rule. The modeling of time series in state space form has advantages over other techniques both in interpretability and estimation. Most of the technical details required in its implementation are given in Chapter 1.

For the CDS component structure, equations (8.12) and (8.13) are estimated to extract information on the unobserved components as well as the unknown parameters of the model.

8.6.3 Linear Regression Model for the Determinants of the CDS Components

We use a regression model to examine the possible relationships between the temporary component and persistent component extracted form the iTraxx indices and the potential drivers, namely, the short-term interest rate, the slope of the yield curve, equity volatility, liquidity (i.e., the bid-ask spread) and credit rating premia.

The specific regression model is as below,

$$\text{dependent} = \alpha + \beta_1 \cdot \text{short rate} + \beta_2 \cdot \text{yield slope}$$
$$+ \beta_3 \cdot \text{volatiltiy} + \beta_4 \cdot \text{spread} + \beta_5 \cdot \text{credit rating} + e$$

where dependent is the time series of interest, for example, 5-year iTraxx Europe or its components, etc and e is the error term in regression model.

All these explanatory variables and their potential influences have been discussed in an intuitive manner in a previous section. In the next section we will discuss the model estimated parameters as well as the nature of the inferred components of the CDS series.

8.7 Analyzing Empirical Results

8.7.1 Model Parameters and the Extracted Components

The adaptive filtering algorithm due to Kalman allows us to estimate the unknown parameters of the model by maximizing the log likelihood function. At the same time it produces optimal inference about the state variables. These filtered state variables are our extracted components consistent with the dynamic specification of the model.

Table 8.5 lists the estimation results for our two component model along with the asymptotic standard errors in parentheses. Firstly, for iTraxx Europe and iTraxx Europe Hivol, all parameter estimates are significant except for the variance of the measurement equation for CDS^5 for iTraxx Europe Hivol. The fact that all the measurement equation error variances are very small is indicative of the appropriateness of the model specification. The estimated values for κ, the speed of mean reversion for the transient component, is 6.45% per annum for iTraxx Europe and 8.791% per annum for iTraxx Europe Hivol. For iTraxx Europe Crossover it is only 0.71%, and is not statistically significant. This means that there is no mean-reversion in the transient component of iTraxx Europe Crossover. It should not be all that surprising for this different behavior as this is the index for sub-investment grade firms. The transient component of the Crossover index demonstrates a more diffusive characteristic.

We now focus on the fact that the estimation results for μ, the drift term in the persistent component, are all negative and statistically significant. Negative drift indicates a downward trend in this component of iTraxx index over the sample period. This is consistent with the original observation of these index time series. The annual drift for iTraxx Europe, iTraxx Europe Hivol and iTraxx Europe Crossover are -0.3360, -0.2599 and -0.3537 respectively. We believe that this downward trend is mainly due to the increasing sophistication of the credit default swap market as well as enhanced liquidity and efficiency. It makes intuitive sense that such a systemic characteristic would be captured by the persistent component of the CDS spreads.

Next, we turn to the diffusion coefficients. The diffusion coefficients for the transient components for iTraxx Europe, iTraxx Europe Hivol and iTraxx Europe Crossover are 0.2413, 0.3085 and 0.3321. The corresponding figures for the persistent components are 0.2527, 0.1741, and 0.2087. We can make some qualitative assessment of the fact that the diffusion coefficient, σ_1, for the transient components for iTraxx Europe Hivol and iTraxx Europe Crossover are higher than that of iTraxx Europe as a whole. This is possibly due to the fact that the credit default swap spreads with underlying entities having lower credit ratings are more sensitive to market news in the short-term. However, from a longer term perspective, the lower credit rating firms have milder variations. The firms that are already on lower credit ratings, have less probability of downgrading, in the longer term, compared with those of higher credit rating firms. So, from this point of view, the diffusion coefficients of the persistent components, σ_2, of iTraxx Europe Hivol and iTraxx Europe Crossover should be lower than that of iTraxx Europe. This is what the estimated results demonstrate.

Furthermore, comparing within each of the three benchmark iTraxx indices, the volatility of the persistent components of iTraxx Europe Hivol and iTraxx Europe Crossover are smaller than that of the respective transient components. The persistent components are less likely to be influenced by market news and events of short-term in nature. The differences in the volatility level of the two components for iTraxx Europe is only marginal. The reason for this is that this series includes more of the firms that belong to higher credit rating categories.

The differences in the pattern of behavior of the two components of the CDS spread series are quite instructive and without such decomposition it would be hard to get such in-depth understanding. Last but not the least; it is interesting to note that all the measurement error variances are quite small indicating the efficacy of the model. Figure 8.7 plots the two extracted components for each of the series against time.

8.7.2 Determinants of the Extracted Components

In this sub-section we analyze the explanatory power of several variables suggested in the literature for the components as well as for the original CDS spread series itself. This helps us understand the differences in the impact these variables have as well as reconcile some of the conflicting results reported in the literature where such a component approach is not implemented. This analysis is carried out in a linear regression framework consistent with the reported studies in this area. In some cases we may have non-stationary series on both sides of the regression equation, but it does not necessarily make the results spurious. As long as the regression residuals are stationary the usual implications of t- and F-tests and R-squares are applicable (see Gujarati (2004), page 822-824).

Table 8.6 lists the estimation results for the explanatory regressions along with the robust standard errors in parentheses below each parameter estimates. The explanatory power of the theoretical variables for levels of default swap spreads are high, approximately 70%, which is consistent with Ericsson et al (2009) and Abid et al (2006a, b). Ericsson et al. (2009) find the explanatory power of the theoretical variables for levels of default swap spreads are approximately 60% and they also point out that the R-square for levels regressions goes up to more than 70% if we add in other explanatory variables as in Collin-Dufresne, Goldstein and Martin (2001).

From Table 8.6 we find that the robust R-square is highest when we regress persistent component against all those explanatory variables. The regression residuals are all found to be stationary. Firstly, the explanatory power of these traditional variables for the persistent component is more than 80% and is much higher than that of the transient component, which is around 50%. Secondly, we notice that since most of the parameters are significant, it implies that the explanatory variables employed here have valuable information for predicting movement in credit default swap spreads. Again, the higher robust R-square seems to show that the traditional explanatory variables can explain more about the persistent component of CDS spreads extracted from the original data relative to the transient component. Next we take a detailed look at these results.

To the extent that the CDS spread indices for two maturities (5-year and 10-year) are valued in the market with respect to the economic fundamentals, we would expect the estimated results for the transient component to be more akin to that of the 5-year maturity index. Similarly, we expect the results for the persistent component to be more like that of the 10-year maturity index.

We can verify this by checking their relative sensitivity to different explanatory variables. For example, iTraxx Europe 5Y is more sensitive to the short-term interest rate than iTraxx Europe 10Y. We would, thus, expect the transient component to be more sensitive to the short-term interest rate than the persistent component, and vice verse. Not surprisingly, the estimated results do support our hypothesis and provide evidence of support for our modeling approach. To be precise, the absolute value of β_1 is larger for all three iTraxx 5Y series than for its counterpart iTraxx 10Y series and it is also larger for the transient component compared to the persistent component. The same pattern of behavior can easily be found for β_3, β_4 and β_5 in the iTraxx Europe and the iTraxx Europe Hivol data. However, it is not so obvious in the iTraxx Europe Crossover case. We will take up this point again later.

Most of our results (from Table 8.6) are consistent with those published in related studies. For example, let us examine the negative relationship between CDS and the short-term interest rate. As we would expect, the short-term interest rate has more impact on the transient component. The result also supports the proposition that the CDS spread is positive correlated with equity market volatility. Again, the corresponding coefficient of equity volatility confirms our hypothesis that this has more influence on the transient component. All the coefficients for the short-term interest rate and equity market volatility are statistically significant at the 1% level of significance.

Our results also suggest that illiquidity may cause higher CDS spread levels which is indicated by a positive coefficient for the liquidity proxy β_4. But when we compare the extent of its impact on the two components, we find that it has more influence on the persistent component implying that the influence of illiquidity is quite persistent. However, for sub-investment grade firms, the influence of this factor does not vary much between the transient and the persistent components.

This indicates that the CDS's whose underlying firms have lower credit ratings are not affected as strongly by its market liquidity as other influential elements. This result can also be verified by comparing the robust R-square for the regression model without a liquidity proxy. After omitting the liquidity proxy from the model for the transient component of iTraxx Europe Crossover, the adjusted R-square only decreases by 0.0016 to 0.7235. But when we omit the other variables, for example, the proxy of credit premia, the adjusted R-square sharply decreases by 0.2378 to 0.4973. We have not presented these results in the tables to conserve space, but they are available on request.

Another set of interesting parameters to examine are β_2 and β_5, the coefficients for the slope of the yield curve and the credit rating premia, respectively. For all three iTraxx benchmark indices, the coefficients of the slope of the yield curve are positive for the persistent component, but negative for the transient component. For each individual index this coefficient is negative. For example, β_2 is 0.1088 and statistically significant for the persistent component of iTraxx Europe, while it is negative for all other three counterpart regressions. That is, β_2 is -0.2666 for the transient component (iTraxx Europe), -0.0918 in iTraxx Europe 5Y and -0.2920 in iTraxx Europe 10Y. As we have explained in an earlier section, the slope of the yield curve is as an indication of overall economic health. Theoretically, an increase in the slope of the yield curve should increase the expected future short-term interest rate and result in a decrease in the CDS spread if the negative relation between credit risk and the short-tem interest rate holds.

But the influence of the slope is quite clear and consistent in our results based on our component methodology. For example, the absolute value of β_2, which can be interpreted as the extent of influence, in iTraxx index regressions are quite close, but are different in the two component-based regressions. The same pattern can be found in all three iTraxx indices. So, the results in this paper provide clear evidence that the slope of the yield curve affects the two components differently. Without these component distinctions the impact of the slope on CDS could turn out conflicting as in Avramov et al. (2007). With a long-term perspective, the increased slope indicating higher expected short-term interest rate may frustrate the market and hence increase the credit default swap spread. This gets reflected via the persistent component. This result also supports the argument that the short interest rate is negatively related to credit default swap spreads and especially to the transient component. We got negative coefficient of slope of yield curve for the transient components.

The credit rating premium is another explanatory variable developed in this study in order to capture the influence of credit rating. The positive coefficient is quite intuitive. This relationship does not appear to be significant for iTraxx Europe 10Y and iTraxx Hivol Europe 10Y. But when we look back to Figure 8.3, we find that the credit rating premia for 10 year maturity contracts are much lower than the credit rating premia for 5 year maturity contracts. As we would expect the importance of credit rating premia (captured by the parameter β_5) increases as more lower credit rated entities are included in the series and in particular for the transient components.

Next let's turn to the explanatory variable spread as well as the Table 8.7. In this table, we list the estimated results for regression models just omitting the credit rating premia. Most parameter estimates are statistically significant under 1% significant level. The important information that can be gleaned from Table 8.7 is that it confirms that credit rating premia are more of concern for the transient component, especially with lower credit rating entities in the CDS index. For example, there is not much deterioration in R-square in case of any of the persistent components. But for iTraxx Crossover the R-square drops from 73.69% (Table 8.6) to 49.99% (Table 8.7) for the transient component.

The results clearly show that the robust R-square without credit rating premia (Table 8.7) does not change much for both the transient and the persistent components of iTraxx Europe. But they do change for the other two iTraxx indices and especially for iTraxx Europe Crossover which is the index with lowest credit rating. So this result confirms the argument that we earlier put forward that for CDS spreads with higher credit rating underlying entities, credit rating has only marginal difference in impact for the two components. While for CDS spreads with lower credit rating underlying entities, the credit rating impact differs depending on the components.

8.8 Concluding Summary

Credit risk analysis is important for valuing corporate bonds, swaps and credit derivatives and plays a critical role in managing the credit risk of bank loan portfolios. Generally it is argued that credit derivatives (especially credit default swaps) are a much better proxy for credit risk since the majority of fundamental variables predicted by credit risk pricing theories have a significant influence on credit default swap prices.

The ideas put forward in this analysis are quite innovative and can provide some hints for later research on the credit derivatives market and the estimated results in this paper are consistent. Firstly, we extract information from the original data using the latent two-factor model used in Schwartz and Smith (2000) and get a temporary component and a persistent component. The intuition behind this is that we extract latent factors which have different movements over time in order to classify different information sets. The results from our latent factor model show that this model is realistic and agrees with our original information. Secondly, we then take a closer look at the different influences of theoretical credit risk explanatory variables on CDS indices and both their temporary and persistent components.

The results from our latent factor model and multi-factor linear regression model show that the temporary component does behave differently from the persistent component and these theoretical explanatory variables do have different influences both in direction and degree on the two components. Also, the regression results show that the theoretical explanatory variables can explain more for the persistent component.

Table 8.1 BBA Credit Derivatives Panel.

Type	2000	2002	2004	2006
Basket products	6.0%	6.0%	4.0%	1.8%
Credit linked notes	10.0%	8.0%	6.0%	3.1%
Credit spread options	5.0%	5.0%	2.0%	1.3%
Equity linked credit products	n/a	n/a	1.0%	0.4%
Full index trades	n/a	n/a	9.0%	30.1%
Single-name credit default swaps	38.0%	45.0%	51.0%	32.9%
Swaptions	n/a	n/a	1.0%	0.8%
Synthetic CDOs – full capital	n/a	n/a	6.0%	3.7%
Synthetic CDOs – partial capital	n/a	n/a	10.0%	12.6%
Trenched index trades	n/a	n/a	2.0%	7.6%
Others	41.0%	36.0%	8.0%	5.7%

Table 8.2 Descriptive Statistics for iTraxx Index Level and Spread Changes.

Maturity	Europe 5Year	10Year	Europe HiVol 5Year	10Year	Europe Crossover 5Year	10Year
Panel A: Statistics for natural logarithm of iTraxx index level						
Min	3.0143	3.6936	3.6784	4.2356	5.0102	5.2941
Mean	3.4898	3.9642	4.1005	4.5124	5.5451	5.7771
Median	3.5660	3.9886	4.1143	4.4906	5.5848	5.7977
Max	4.0962	4.4145	4.7465	4.9978	6.0970	6.2222
StDev	0.2090	0.1329	0.2081	0.1429	0.1686	0.1320
Skewness	-0.3613	-0.1273	0.0229	0.4388	-0.4760	-0.8286
Kurtosis	-0.5593	-0.4042	-0.7839	-0.5470	0.5606	1.9650
ADF Test	0.6900	0.4700	0.4700	0.2700	0.1600	0.0900
PP test	0.7000	0.5500	0.5100	0.3500	0.1900	0.1200
Panel B: Statistics for the first difference of natural logarithm of iTraxx index level						
Min*100	-11.1463	-8.7046	-11.6056	-9.7667	-9.3260	-7.8149
Mean*100	-0.1029	-0.0451	-0.0870	-0.0280	-0.0601	-0.0077
Median*100	-0.1344	-0.1115	-0.2500	-0.1471	-0.2317	-0.1669
Max*100	19.3867	12.8844	22.7082	16.3525	32.1692	26.8990
StDev	0.0199	0.0143	0.0230	0.0183	0.0257	0.0203
Skewness	1.43822	0.9745	2.3215	2.1637	4.0378	4.2130
Kurtosis	16.3768	12.4792	21.1226	22.3978	40.3180	46.9672
ADF Test	0.0000	0.0000	0.0000	0.0000	0.0000	0.0000
PP test	0.0000	0.0000	0.0000	0.0000	0.0000	0.0000
Panel C: Descriptive Statistics for log return on MSCI						
Min*100	-3.1640					
Mean*100	0.0806					
Median*100	0.1110					
Max*100	2.5108					
StDev	0.0076					
Skewness	-0.4333					
Kurtosis	1.6111					

MacKinnon approximate p-value is listed in row of Augmented Dickey-Fuller test (ADF Test) and Phillips-Perron test (PP Test) with unit root test model without trend and with 4 lags.

Table 8.3 Numerical Value of Credit Rating.

Moody's	S&P	Numerical Value
Aaa	AAA	1
Aa1	AA+	2
Aa2	AA	3
Aa3	AA-	4
A1	A+	5
A2	A	6
A3	A-	7
Baa1	BBB+	8
Baa2	BBB	9
Baa3	BBB-	10
Ba1	BB+	11
Ba2	BB	12
Ba3	BB-	13
B1	B+	14
B2	B	15
B3	B-	16
Caa1	CCC+	17
Caa2	CCC	18
Caa3	CCC-	19

Table 8.4(a) Component Names in Series 7 of iTraxx Europe.

No.	Ticker	Rating	Value	No.	Ticker	Rating	Value
1	AAB	Aa2	3	16	CMZB	Aa3	4
2	AEGON	A2	6	17	DB	Aa1	2
3	ALZ	Aa3	4	18	HANRUE	A3	7
4	ASSGEN	Aa3	4	19	ISPIM	Aa2	3
5	AVLN	A3	7	20	MUNRE	Aa3	4
6	AXASA	A2	6	21	RSA	A3	7
7	MONTE	Aa3	4	22	SCHREI	Aa2	3
8	BPIIM	A2	6	23	RBS	Aaa	1
9	BBVASM	Aa1	2	24	CRDIT	Aa2	3
10	BCPN	Aa3	4	25	ZURNVX	A1	5
11	BESNN	Aa3	4	26	VLVY	A3	7
12	SANTAN	Aa1	2	27	BMW	A1	5
13	BACR	Aa1	2	28	MICH	Baa2	9
14	BNP	Aa1	2	29	CONTI	Baa1	8
15	CAPIM	A1	5	30	DCX	Baa1	8

Table 8.4(a) Component Names in Series 7 of iTraxx Europe (Contd.).

No.	Ticker	Rating	Value	No.	Ticker	Rating	Value
31	GKNLN	Baa3	10	71	ENEL	A1	5
32	PEUGOT	Baa1	8	72	ELEPOR	A2	6
33	RENAUL	Baa1	8	73	ENBW	A2	6
34	VLOF	Baa2	9	74	FRTUM	A2	6
35	VW	A3	7	75	GASSM	A1	5
36	ACCOR	BBB	9	76	GAZDF	Aa1	2
37	ELTLX	Baa2	9	77	IBERDU	A2	6
38	ABLN	Baa2	9	78	NGGLN	Baa1	8
39	ALTSM	Baa2	9	79	REPSM	Baa1	8
40	BATSLN	Baa1	8	80	RWE	A1	5
41	CBRY	Baa2	9	81	LYOE	A2	6
42	CARR	A2	6	82	UNFSM	Baa1	8
43	COFP	BBB-	10	83	UU	A3	7
44	CPGLN	Baa2	9	84	VATFAL	A2	6
45	LUFTHA	Baa3	10	85	VIEFP	A3	7
46	DIAG	A3	7	86	ADO	Baa2	9
47	DSGILN	Baa2	9	87	AKZO	A3	7
48	GUSLN	Baa1	8	88	LORFP	Baa3	10
49	GLHLN	A2	6	89	BYIF	A3	7
50	AUCHAN	A	6	90	CIBASC	Baa2	9
51	HENKEL	A2	6	91	SGOFP	Baa1	8
52	IMPTOB	Baa3	10	92	EADFP	A1	5
53	KINGFI	Baa3	10	93	GLENCR	Baa3	10
54	PHG	A3	7	94	HANSON	Baa3	10
55	MOET	A-	7	95	ICI	Baa2	9
56	MKS	Baa2	9	96	DSM	A2	6
57	METFNL	Baa2	9	97	LAFCP	Baa2	9
58	PRTP	BBB-	10	98	LINGR	Baa1	8
59	MRWLN	Baa2	9	99	SANFP	A1	5
60	EXHO	BBB+	9	100	SIEM	Aa3	4
61	SCACAP	Baa1	8	101	SOLBBB	A2	6
62	TATELN	Baa2	9	102	STORA	Baa3	10
63	TSCO	A1	5	103	TKAGR	Baa2	9
64	TMMFP	Baa2	9	104	UPMKYM	Baa2	9
65	ULVR	A1	5	105	VINCI	Baa1	8
66	CENTRI	A3	7	106	BERTEL	Baa1	8
67	EOAGR	A2	6	107	BRITEL	Baa1	8
68	EDNIM	Baa2	9	108	DT	A3	7
69	EDF	Aa1	2	109	FRTEL	A3	7
70	ELESM	A3	7	110	OTE	Baa1	8

Table 8.4(a) Component Names in Series 7 of iTraxx Europe (Contd.).

No.	Ticker	Rating	Value	No.	Ticker	Rating	Value
111	KPN	Baa2	9	121	TLIASS	A2	6
112	PSON	Baa1	8	122	VIVFP	Baa2	9
113	PUBFP	Baa2	9	123	VOD	Baa1	8
114	REEDLN	Baa1	8	124	WOLKLU	Baa1	8
115	RTRGRP	Baa1	8	125	WPPLN	Baa2	9
116	STM	A3	7				
117	TITIM	Baa2	9				
118	TELEFO	Baa1	8				
119	TKA	A3	7				
120	TELNOR	A2	6	Average Rating		A2/A3	6.896

Table 8.4(b) Component Names in Series 7 of iTraxx Europe Crossover.

No.	Ticker	Rating	Value	No.	Ticker	Rating	Value
1	ALUFP	Ba2	12	26	KABEL	Ba3	13
2	NELL	Ba3	13	27	AHOLD	Baa3	10
3	BCMAU	B1	14	28	LADLN	Ba2	12
4	BAB	Baa3	10	29	MESSA	B3	16
5	CWLN	B1	14	30	TDCDC	Ba3	13
6	CAPP	BB+	11	31	NSINO	Ba1	11
7	CODERE	B2	15	32	NXPBV	B1	14
8	COGNIS	B2	15	33	ONOFIN	B3	15
9	COLTLN	B2	15	34	PORTEL	Baa2	9
10	CORUS	Ba1	11	35	PROSIE	Ba1	11
11	DEGUSS	Baa3	10	36	RALFP	Ba1	11
12	EMI	B1	14	37	RAYAC	Ba1	11
13	FIAT	Ba1	11	38	RHA	Ba3	13
14	FKI	Ba2	12	39	SAS	B1	14
15	FREGR	Ba2	12	40	SEAT	Ba3	13
16	SDC	B2	16	41	MDPAC	B2	16
17	GFCFP	Baa3	10	42	SOLSM	Baa3	10
18	GROHE	B2	16	43	VNU	B2	16
19	HAVAS	Ba2	12	44	RNK	Ba3	13
20	HELLAS	B2	16	45	TUIGR	B1	14
21	INEGRP	B1	14	46	IESYRP	B3	15
22	IFX	B1	14	47	UPC	B1	14
23	IPRLN	Ba3	13	48	VMED	B2	16
24	ISYSLN	Ba3	13	49	WDAC	B2	16
25	ITVLN	Baa3	10	50	WINDIM	B2	16
				Average Rating		Ba3/B1	13.1

Table 8.4(c) Component Names in Series 7 of iTraxx Europe Hivol.

No.	Ticker	Rating	Value	No.	Ticker	Rating	Value
1	ACCOR	BBB	9	16	ICI	Baa2	9
2	ELTLX	Baa2	9	17	KINGFI	Baa3	10
3	BRITEL	Baa1	8	18	KPN	A2	6
4	CBRY	Baa2	9	19	MKS	Baa2	9
5	COFP	BBB-	10	20	PSON	Baa1	8
6	CIBASC	Baa2	9	21	PRTP	BBB-	10
7	CPGLN	Baa2	9	22	MRWLN	Baa2	9
8	CONTI	Baa1	8	23	STORA	Baa3	10
9	DCX	Baa1	8	24	TATELN	Baa2	9
10	LUFTHA	Baa3	10	25	TITIM	Baa2	9
11	DSGILN	Baa2	9	26	TMMFP	Baa2	9
12	GUSLN	Baa1	8	27	UPMKYM	Baa2	9
13	GKNLN	Baa3	10	28	VLOF	Baa2	9
14	GLENCR	Baa3	10	29	VIVFP	Baa2	9
15	HANSON	Baa3	10	30	WOLKLU	Baa1	8
				Average Rating Baa1/Baa2 8.97			

Table 8.5 Estimated Results for Schwartz and Smith (2000) Two-Factor Model.

	iTraxx Europe	iTraxx Hivol	iTraxx Crossover
k	0.0645	0.08791	0.0071
	(1.0558e-05)	(3.4554e-10)	(0.1106)
μ	-0.3360	-0.2599	-0.3537
	(2.5209e-06)	(1.1254e-10)	(0.0259)
σ_1	0.2413	0.3085	0.3321
	(0.0004)	(3.9884e-08)	(0.0019)
σ_2	0.2527	0.1741	0.2087
	(0.0001)	(1.2685e-07)	(0.0009)
d	0.3792	0.1382	0.4155
	(4.6337e-05)	(4.4952e-06)	(0.0076)
ρ	0.1322	0.5954	0.5064
	(0.0003)	(2.7771e-10)	(0.0126)
a	1.8004	2.3572	3.72658
	(1.4920e-05)	(1.4121e-10)	(6.99288e-06)
b	2.6774	3.4004	4.34888
	(5.8563e-06)	(4.6739e-06)	(0.00788)
$\sigma_{1,\varepsilon}$	7.3658e-06	1.0000e-10	2.4998e-05
	(6.0727e-11)	(3.0614e-10)	(1.3065e-10)
$\sigma_{2,\varepsilon}$	1.0000e-10	2.8374e-05	2.8588e-06
	(2.5993e-11)	(2.2415e-10)	(3.3042e-11)
Log-likelihood	4654.8879	4315.0815	4379.8852

Asymptotic estimator of standard error is reported below each estimated parameter. Here we adopt BHHH estimator as variance estimators for an MLE. More detail information may be found in William H. Greene, "Econometric Analysis" fifth edition, Page 480-482.

Table 8.6 Estimation Results for Regression Model.

	5Y	10Y	Temporary	Persistent
Panel a: Dependent variable: iTraxx Europe				
α	3.9012***	4.8587***	1.6975***	0.4894***
	(0.1128)	(0.0869)	(0.0996)	(0.0450)
β_1	-1.0319***	-0.9769***	-0.7618***	-0.3982***
	(0.0374)	(0.0346)	(0.0337)	(0.0207)
β_2	-0.0918***	-0.2920***	-0.2666***	0.1088***
	(0.0178)	(0.0199)	(0.0188)	(0.0117)
β_3	21.3165***	16.2558***	14.1927***	6.2780***
	(2.4143)	(2.7215)	(2.7101)	(0.7138)
β_4	0.0757***	0.1258***	0.0299**	0.1106***
	(0.0128)	(0.0196)	(0.0114)	(0.0148)
β_5	0.2596***	0.0651	0.0890**	0.2196***
	(0.0522)	(0.0452)	(0.0422)	(0.0223)
R^2	0.8405	0.6684	0.4413	0.9256
Panel b: Dependent variable: iTraxx Hivol				
α	4.7110***	5.3324***	1.7042***	0.6155***
	(0.1627)	(0.1016)	(0.1265)	(0.0465)
β_1	-1.5296***	-0.9635***	-1.1917***	-0.3127***
	(0.0489)	(0.0454)	(0.0396)	(0.0227)
β_2	-0.3439***	-0.4009***	-0.4086***	0.0852***
	(0.0326)	(0.0174)	(0.0254)	(0.0084)
β_3	29.1548***	17.6193***	20.1077***	9.6246***
	(2.7971)	(2.3884)	(2.4564)	(0.7105)
β_4	0.0403	0.2396***	0.0085	0.0176*
	(0.0254)	(0.0286)	(0.0171)	(0.0091)
β_5	0.4387***	0.0631	0.2942***	0.1498***
	(0.0675)	(0.0496)	(0.0515)	(0.0258)
R^2	0.7038	0.6338	0.5335	0.8640

Table 8.6 Estimation Results for Regression Model (Contd.).

	5Y	10Y	Temporary	Persistent
Panel c: Dependent variable: iTraxx Europe Crossover				
α	3.6103^{***}	4.252241^{***}	-0.0872	0.0936
	(0.1133)	(0.120182)	(0.0921)	(0.0677)
β_1	-0.9257^{***}	-0.6281^{***}	-0.5614^{***}	-0.2419^{***}
	(0.0389)	(0.0464)	(0.0300)	(0.0309)
β_2	-0.1017^{***}	-0.1620^{***}	-0.2707^{***}	0.1794^{***}
	(0.0187)	(0.0178)	(0.0167)	(0.0117)
β_3	21.1169^{***}	20.3619^{***}	10.2978^{***}	13.7750^{***}
	(2.5014)	(2.6043)	(1.9587)	(1.1726)
β_4	0.1136^{***}	0.1169^{***}	0.0793^{***}	0.0716^{***}
	(0.0180)	(0.0173)	(0.0174)	(0.0074)
β_5	1.2648^{***}	1.0221^{***}	0.8761^{***}	0.2333^{***}
	(0.0439)	(0.0454)	(0.0361)	(0.0346)
R^2	0.7645	0.6807	0.7369	0.8637

The model provided in this table is

$$dependent = \alpha + \beta_1 \cdot riskfree\ rate + \beta_2 \cdot yield\ slope + \beta_3 \cdot volatiltiy$$
$$+ \beta_4 \cdot spread + \beta_5 \cdot credit\ rating + \varepsilon .$$

We use Stata command "regress, robust" and the robust standard error is proved in parentheses below each estimator and robust R^2 is provided as well.

1%, 5% and 10% significance levels are indicated by ***, **, and *, respectively.

Table 8.7 Estimation Results for Regression Model.

	iTraxx Europe		iTraxx Europe Hivol		iTraxx Europe Crossover	
	Temporary	Persistent	Temporary	Persistent	Temporary	Persistent
α	1.8728^{***}	0.8769^{***}	2.3107^{***}	0.8375^{***}	1.4836^{***}	0.4062^{***}
	0.0367	0.0268	0.0589	0.0259	0.0670	0.0436
β_1	-0.7402^{***}	-0.3677^{***}	-1.1573^{***}	-0.2648^{***}	-0.3362^{***}	-0.1601^{***}
	0.0354	0.0212	0.0425	0.0211	0.0460	0.0289
β_2	-0.2797^{***}	0.0729^{***}	-0.4676^{***}	0.0693^{***}	-0.4066^{***}	0.1597^{***}
	0.0157	0.0126	0.0264	0.0085	0.0200	0.0116
β_3	14.4515^{***}	6.9112^{***}	21.5621^{***}	9.7720^{***}	14.6423^{***}	14.9572^{***}
	2.7329	0.7847	2.6707	0.7324	2.8155	1.3038
β_4	0.0390^{***}	0.1252^{***}	0.0312	0.0385^{***}	0.1415^{***}	0.0894^{***}
	0.0118	0.0166	0.0246	0.0097	0.0265	0.0085
R^2	0.4369	0.9142	0.5048	0.8557	0.4999	0.8523

The model provided in this table is
$$\text{dependent} = \alpha + \beta_1 \cdot \text{riskfree rate} + \beta_2 \cdot \text{yield slope} + \beta_3 \cdot \text{volatiltiy} + \beta_4 \cdot \text{spread} + \varepsilon.$$

We use Stata command "regress, robust" and the robust standard error is proved in parentheses below each estimator and robust R2 is provided as well.

1%, 5% and 10% significance levels are indicated by ***, **, and *, respectively.

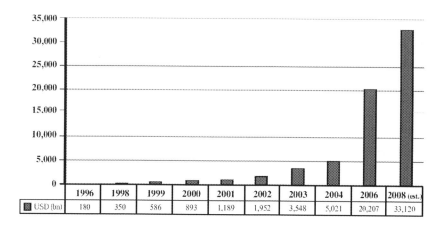

USD (bn)	1996	1998	1999	2000	2001	2002	2003	2004	2006	2008 (est.)
	180	350	586	893	1,189	1,952	3,548	5,021	20,207	33,120

Figure 8.1 Global Credit Derivatives Market $billion (Source: British Bankers'
Association, Credit Derivatives Report 2006).

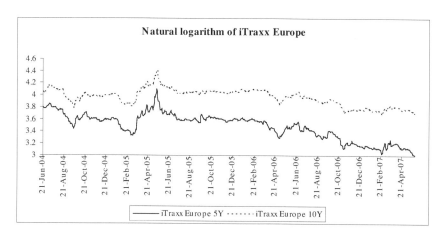

Figure 8.2 (Panel A) iTraxx Europe Index.

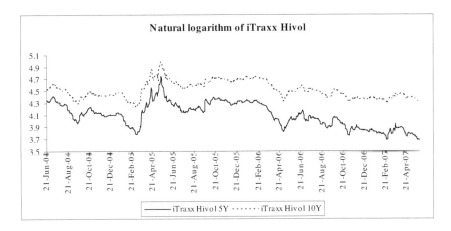

Figure 8.2 (Panel B) iTraxx Europe Hivol Index.

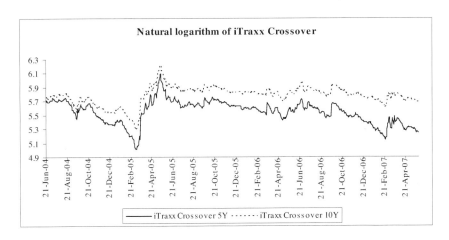

Figure 8.2 (Panel C) iTraxx Europe Crossover Index.

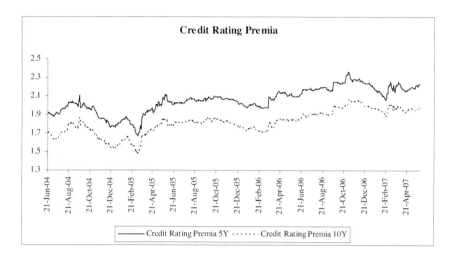

Figure 8.3 Three Credit Rating Premia for Two Different Maturities.

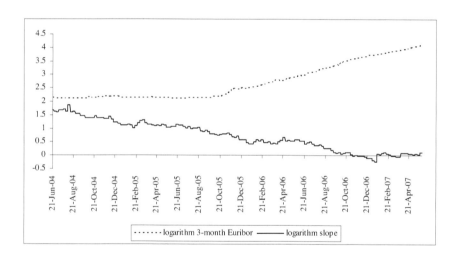

Figure 8.4 Three-Months Euribor and Slope of Yield Curve.

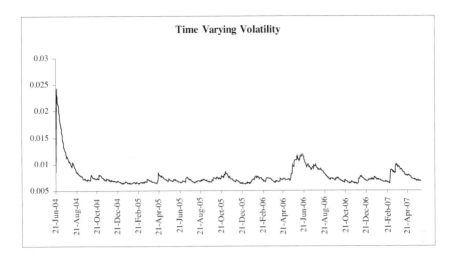

Figure 8.5 European Equity Market Volatility (GARCH (1,1) type).

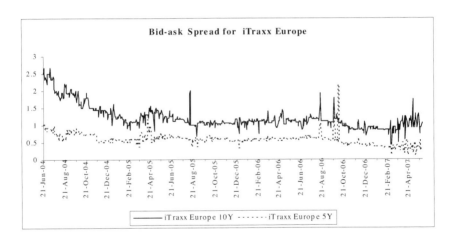

Figure 8.6 (Panel A) Bid-ask spread for iTraxx Europe index.

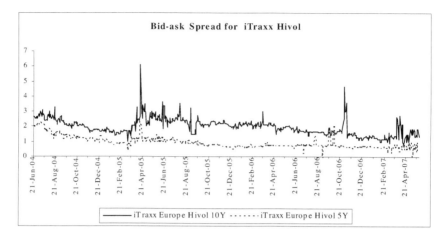

Figure 8.6 (Panel B) Bid-ask spread for iTraxx Europe Hivol index.

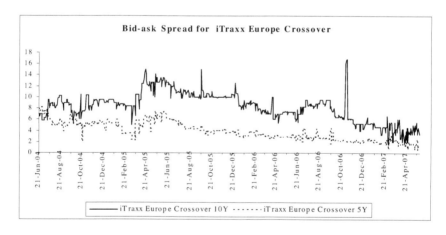

Figure 8.6 (Panel C) Bid-ask spread for iTraxx Europe Crossover index.

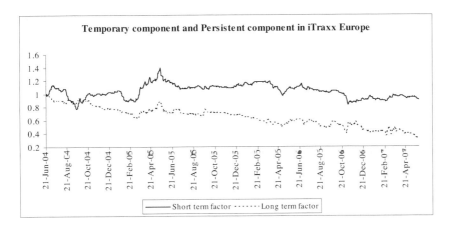

Figure 8.7 (Panel A) Temporary component and Persistent components of iTraxx Europe index.

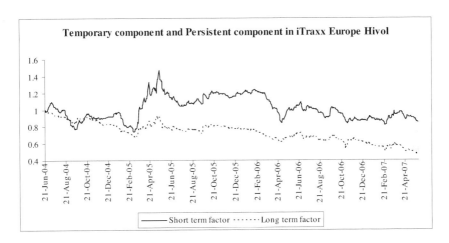

Figure 8.7 (Panel B) Temporary component and Persistent components of iTraxx Europe Hivol index.

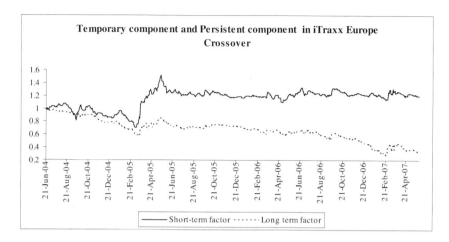

Figure 8.7 (Panel B) Temporary component and Persistent components of iTraxx Europe Crossover index.

Chapter 9

CDS Options, Implied Volatility and Unscented Kalman Filter

Implied volatility captures market's assessment of future uncertainty as impounded in the prices of derivatives and contains reliable information about the subsequent realized volatility. If so, then time-variation in implied volatility would also reflect time-variation in the uncertainty about future volatility. A higher implied volatility may be associated with higher economic-state uncertainty. These arguments suggest that the implied volatility is a better proxy to catch the perceived credit risk and uncertainty in the future. Models of implied volatilities in the Black-Scholes framework implicitly assume Gaussian distribution. This chapter focuses on Heston (1993) framework to allow non-Gaussian distribution in the process. By allowing the forward credit default swap rate to be a state variable, it is shown how Heston model may be estimated by stochastic filtering. This, however, requires Unscented Kalman Filter since the assumptions of linear Kalman filter are violated.

This chapter is to explain how to apply a state-space approach and unscented Kalman Filter to estimate the implied volatility from market price of credit default swap (CDS) option based on Heston model.

The empirical analysis of implied volatility from the credit market is important on several dimensions, particularly when it is compared with those of the equity market. First, most researchers focus on the daily changes in prices from these two markets, but because of nonlinear relationship, we should also examine the relationship using the second moments. Second, Ross (1989) indicates that the volatility is more important than mean price changes in the transmission of information. Third, instead of using realized volatility, the implied volatility is expected to capture market's assessment of future uncertainty as impounded in the prices of derivatives.

This chapter has benefited immensely from the work done by Peipei Wang, my PhD student prior to writing this book.

9.1 Background to Stochastic Volatility

Implied volatility is expected to capture market's assessment of future uncertainty as impounded in the prices of derivatives and contains reliable information about the subsequent realized volatility (Blair et al. (2001); Christensen and Prabhala (1998)). Coval and Shumway (2001) and Bakshi and Kapadia (2003) suggest that option prices may contain a premium attributed to stochastic volatility. If so, then time-variation in implied volatility would also reflect time-variation in the uncertainty about future volatility. A higher implied volatility may be associated with higher economic-state uncertainty. All these arguments suggest that the implied volatility is a better proxy to capture the perceived credit risk and uncertainty in the future.

There are several economic, empirical, and mathematical reasons for choosing a model with such a form (Cont (2001)). Firstly, the Black-Scholes model assumes the implied volatilities for different maturity options are deterministic, which is not appropriate for CDS option. Secondly, empirical studies have shown that an asset's log-return distribution is non-Gaussian and Heston model can imply a number of different distributions.

Generally, the calibration of Heston model can rely on the approach by Bakshi et al. (1997), which means each day, we need to collect N option prices and the number of option prices and this number must be greater than or equal to the number of parameters in the model to avoid underdetermined systems. It is difficult to implement sometimes because of insufficient number of option prices.

In this chapter, we calibrate the model within state space framework and then employ Unscented Kalman Filter (UKF) to estimate parameters in this model. The idea behind this method is that, we notice that the volatility of underlying assets is a key determinant for option price. In the context of state space, the volatility of forward credit default swap rate is a state variable, we say, the dynamic of option price is driven by this state variable.

9.2 Heston Model in Brief

Within Heston (1993) framework, let's assume that the spot credit default swap (CDS) spread at time t, $S(t)$ follows the diffusion

$$dS(t) = \mu S dt + \sqrt{V(t)} S dZ_1(t) \qquad (9.1)$$

and the variance $V(t)$ follows the process,

$$dV(t) = \left[\kappa_v \left(\bar{v} - V(t) \right) \right] dt + \sigma_v \sqrt{V(t)} dZ_2(t) \qquad (9.2)$$

where,

κ_v is the speed of mean reversion;

$\bar{v}$ is the long run mean of the variance process;

σ_v is the volatility of the variance process;

dZ_1 and dZ_2 are two standard Wiener processes with correlation ρ.

Let's assume s is forward credit default swap spread and let $Y = \ln(s)$, then by Ito's lemma under the risk-neutral measure Q:

$$dY_t = -\tfrac{1}{2} V_t dt + \sqrt{V_t} dZ_1(t), \qquad (9.3)$$

$$dV_t = \left[\kappa_v (\bar{v} - V_t) \right] dt + \sqrt{V_t} \sigma_v dZ_2(t), \qquad (9.4)$$

$$dZ_1 dZ_2 = \rho dt. \qquad (9.5)$$

Parameter ρ can be interpreted as the correlation between the log-return and volatility of the asset, affects the heaviness of the tail. Parameter σ_v affects the kurtosis of the distribution. When σ_v is 0, the volatility is deterministic and hence log-return will be normally distributed. Increasing σ_v will only increase the kurtosis, creating heavy tails on both sides, which means that the market has a greater chance of extreme movements, so writers of puts must charge more and while those of calls less for a given strike.

The calibration problem has a slight simplification when we price options under risk-neutral measure Q. Evaluation under an equivalent martingale measure (EMM) effectively reduces the number of estimated parameters to five. The following must therefore hold for an option:

$$\text{Value}^P (\kappa_v; \overline{v}; \sigma_v; V_0; \rho; \lambda) = \text{Value}^Q (\kappa_v^*; \overline{v}^*; \sigma_v; V_0; \rho; 0)$$

where

λ is market price of volatility risk;

$\kappa_v^* = \kappa_v + \lambda$,

$\overline{v}^* = \kappa_v \overline{v} / (\kappa_v + \lambda)$.

9.3 State Space Framework

9.3.1 Transition Equation

In Heston (1993) model, the time-varying variance process, equation (9.4) is the CIR process (Cox, Ingersoll, and Ross (1985)). In the risk-neutralized pricing probabilities, the variance $V(t)$ follows the process,

$$dV(t) = \left[\kappa_v^* (\overline{v}^* - V(t)) \right] dt + \sigma_v \sqrt{V(t)} dZ_2(t). \qquad (9.6)$$

Then it's exact transition density of the time-varying variance is a non-central chi-square, $\chi^2 [2cs_t; 2q+2, 2u]$, with $2q+2$ degrees of freedom and parameter of non-centrality $2u$. The probability density of the variance at time t, conditional on its value at the time $t-1$ is given by,

$$f(V_t, t; V_{t-1}, t-1) = ce^{-u-v} \left(\frac{v}{u} \right)^{q/2} I_q \left(2(uv)^{1/2} \right) \qquad (9.7)$$

where,

$c = \dfrac{2\kappa_v^*}{\sigma_v^2 \left(1 - e^{-\kappa_v^* \Delta t}\right)}$, $u = cV_{t-1} e^{-\kappa_v^* \Delta t}$, $v = cV_t$, $q = \dfrac{2\kappa_v^* \overline{v}^*}{\sigma_v^2} - 1$, and $I_q(.)$ is

the modified Bessel function of the first kind with order q.

So the mean and the variance of V_t conditional on $\mathcal{F}_{t-1}$ are given by

$$E(V_t | \mathcal{F}_{t-1}) = V_{t-1} \exp(-\kappa_v^* \Delta t) + \overline{v}^* [1 - \exp(-\kappa_v^* \Delta t)] \qquad (9.8)$$

$$Q_t \equiv Var(V_t|\mathcal{F}_{t-1}) = V_{t-1}(\frac{\sigma_v^2}{\kappa_v^*})[\exp(-\kappa_v^*\Delta t) - \exp(-2\kappa_v^*\Delta t)]$$

$$+ \bar{v}^* \left(\frac{\sigma_v^2}{2\kappa_v^*}\right)[1 - \exp(-\kappa_v^*\Delta t)]^2 .$$

(9.9)

Thus we can get the transition equation:

$$V_t = d + FV_{t-1} + u_t$$

(9.10)

where,

$$E(u_t|\mathcal{F}_{t-1}) = 0 ,$$

(9.11)

$$var(u_t|\mathcal{F}_{t-1}) = Q_t ,$$

(9.12)

$$d = \bar{v}^* \left[1 - \exp(-\kappa_v^*\Delta t)\right] ,$$

(9.13)

$$F = \exp(-\kappa_v^*\Delta t) .$$

(9.14)

9.3.2 Measurement Equation: CDS Option Price

Consider the option for a protection buyer to enter a CDS at a future time $T_a > 0$, $T_a < T_b$, paying a fixed premium rate K at times $T_{a+1},...,T_b$ or until default, in exchange for a protection payment L_{GD} against possible default in $[T_a, T_b]$. Then the CDS option's discounted payoff at time t is

$$\Pi_{a,b}^{call}(t;K) = D_{t,T_a}\left[\frac{CDS_{a,b}(T_a,R_{a,b}(T_a),L_{GD}) -}{CDS_{a,b}(T_a,K,L_{GD})}\right]^+$$

(9.15)

where

$$D_{t,T} = \exp\left(-\int_t^T r_s ds\right), r_s \text{ is stochastic interest rate;}$$

$$L_{GD} \text{ is loss given default;}$$

$$CDS_{a,b}(t,R,L_{GD}) \text{ is the price of standard CDS at time } t .$$

This option exchanges payment rates R at time $T_{a+1},...,T_b$ or until default in exchange for a single protection payment L_{GD}.

From the equation (21.20) in Brigo and Mercurio (2006),

$$CDS_{a,b}\left(T_a,R,L_{GD}\right) = \frac{1_{\{\tau > T_a\}}}{Q\left(\tau > T_a \big| \mathcal{F}_{T_a}\right)}$$

$$\left\{ R \cdot E\left[D(T_a,\tau)\left(\tau - T_{\beta(\tau)-1}\right) 1_{\{T_a < \tau < T_b\}} \Big| \mathcal{F}_{T_a} \right] \right.$$

$$+ R \sum_{i=a+1}^{b} \alpha_i \overline{P}(T_a,T_i) Q\left(\tau > T_a \big| \mathcal{F}_{T_a}\right)$$

$$\left. - L_{GD} E\left[1_{\{T_a < \tau \le T_b\}} D(T_a,\tau) \Big| \mathcal{F}_{T_a} \right] \right\} \qquad (9.16)$$

where,

α_i is the year fraction between T_{i-1} and T_i ;

$T_{\beta(t)}$ is the first date among the T_i's that follows t ;

$$1_{\{\tau > t\}} \overline{P}(t,T) = \frac{1_{\{\tau > t\}}}{Q\left(\tau > t \big| \mathcal{F}_t\right)} E\left[D(t,T) 1_{\{\tau > T\}} \big| \mathcal{F}_t \right].$$

Hence,

$$\Pi_{a,b}^{call}\left(T_a;K\right) = \frac{1_{\{\tau > T_a\}}}{Q\left(\tau > T_a \big| \mathcal{F}_{T_a}\right)}$$

$$\times \left\{ E\left[D(T_a,\tau)\left(\tau - T_{\beta(\tau)-1}\right) 1_{\{T_a < \tau < T_b\}} \Big| \mathcal{F}_{T_a} \right] \right.$$

$$\left. + \sum_{i=a+1}^{b} \alpha_i \overline{P}(T_a,T_i) Q\left(\tau > T_a \big| \mathcal{F}_{T_a}\right) \right\} \left(R_{a,b}(T_a) - K \right)^+ . \quad (9.17)$$

If we ignore the accrued interest term, this becomes

$$\Pi_{a,b}^{call}\left(T_a;K\right) = \frac{1_{\{\tau > T_a\}}}{Q\left(\tau > T_a \big| \mathcal{F}_{T_a}\right)}$$

$$\times \sum_{i=a+1}^{b} \alpha_i \overline{P}(T_a,T_i) Q\left(\tau > T_a \big| \mathcal{F}_{T_a}\right) \left(R_{a,b}(T_a) - K \right)^+ \quad (9.18)$$

$$\equiv 1_{\{\tau > T_a\}} \sum_{i=a+1}^{b} \alpha_i \overline{P}(T_a,T_i) \left(R_{a,b}(T_a) - K \right)^+ .$$

This result makes no assumptions about the distribution of the CDS rate

$$R_{a,b}(t).$$

This formula is very similar to the standard market model for valuing European Swaption. The only difference is that for swaptions, we discount at the risk-free rate, while for CDS options we discount at the risky interest rate. We can now make essentially any assumption we like about the distribution of the CDS rate.

9.3.3 Measurement Equation Derivation

The price of a European call option is given by:

$$\Pi_{a,b}^{call}(t;K) = \sum_{i=a+1}^{b} \alpha_i \overline{P}(t,T_i)\left[R_{a,b}(t)\cdot\Pi_1 - K\cdot\Pi_2\right]. \qquad (9.19)$$

For notational consistency in Heston (1993) model, we write:

$$R_{a,b} \equiv s, \ T_a \equiv T,$$

then

$$\begin{aligned}\Pi_2 &= P\{s_T > K \big| s_t, V_t\} \\ &= P\{\ln(s_T) > \ln(K)\big| s_t, V_t\} \\ &= \frac{1}{2} + \frac{1}{\pi}\int_0^\infty \mathrm{Re}\left(\frac{e^{-iu\ln K}}{iu}\phi(Y_t,V_t,t;iu)\right)du.\end{aligned}$$

Therefore,

$$\Pi_2 = \frac{1}{2} + \frac{1}{\pi}\int_0^\infty \mathrm{Re}\left(\frac{e^{-iu\ln(s_t/K)}}{iu}\hat{\phi}(V_t,t;iu)\right)du, \qquad (9.20)$$

and,

$$\begin{aligned}\Pi_1 &= \frac{1}{2} + \frac{1}{\pi}\int_0^\infty \mathrm{Re}\left(\frac{e^{-iu\ln K}}{iu}e^{-Y_t}\phi(Y_t,V_t,t;1+iu)\right)du \\ &= \frac{1}{2} + \frac{1}{\pi}\int_0^\infty \mathrm{Re}\left(\frac{e^{-iu\ln(s_t/K)}}{iu}\hat{\phi}(V_t,t;1+iu)\right)du \qquad (9.21)\end{aligned}$$

where,

$$\phi(Y_t, V_t, t; u) := E\left[e^{uY_T} \middle| \mathcal{F}_t\right]$$
$$= e^{\alpha(t) + \beta_1(t)Y_t + \beta_2(t)V_t}, \tag{9.22}$$

is the (risk-neutral) characteristic function and $\alpha(t)$, $\beta_1(t)$ and $\beta_2(t)$ are independent from Y_t and V_t.

$$\hat{\phi}(V_t, t; u) := E\left[e^{u(Y_T - Y_t)} \middle| \mathcal{F}_t\right]$$
$$= e^{\alpha(t) + \beta_2(t)V_t}. \tag{9.23}$$

Singleton et al (2000) show that,

$$\alpha(t) = \kappa_v^* \bar{v}^* \int_t^T \beta_2(s) ds$$

$$\beta_1(t) = u$$

$$\beta_2(t) = \frac{-a\left(1 - e^{-\gamma\tau}\right)}{2\gamma - (\gamma + b)\left(1 - e^{-\gamma\tau}\right)}$$

where,

$$\tau = (T - t), b = \sigma_v \rho u - \kappa_v^*, a = u(1 - u), \gamma = \sqrt{b^2 + a\sigma_v^2}.$$

So the measurement equation is nonlinear equation and we can simply express it as,

$$\Pi_{a,b}^{call}(t; K) = h(V_t; \Theta) + \varepsilon_t. \tag{9.24}$$

Finally we get State space model as,

$$\Pi_{a,b}^{call}(t; K) = h(V_t; \Theta) + \varepsilon_t, \ E(\varepsilon_t | \mathcal{F}_{t-1}) = 0, \ var(\varepsilon_t | \mathcal{F}_{t-1}) = R, \tag{9.25}$$

$$V_t = d + FV_{t-1} + \zeta_t, \ E(\zeta_t | \mathcal{F}_{t-1}) = 0, \ var(\zeta_t | \mathcal{F}_{t-1}) = Q_t, \tag{9.26}$$

where,

    $h(.)$ is nonlinear pricing function of option ;

    ε_t is measurement noise, white noise with $var(\varepsilon_t | \mathcal{F}_{t-1}) = R$;

    ζ_t is process noise;

    d and F are defined as above.

9.4 General State Space Model and Filter Revisited

A general state-space model (neglecting control inputs for the sake of clarity) comprise a state transition model which follows a first order Markov process and a measurement model which links actual observations with latent variables.

Measurement equation:

$$y(t) = h\left[x(t), e(t)\right], \qquad (9.27)$$

Transition equation:

$$x(t+1) = f\left[x(t), w(t+1)\right], \qquad (9.28)$$

where

$x(t)$ is called state vector and represent properties that are of interest to us. The process of extracting data from a measured signal, using the state space model is all about estimating the states;

$f(\cdot)$ is a function that describes how the states change during a certain interval;

$w(t)$ is process noise;

$y(t)$ is the observations, or measurements at time t ;

$h(\cdot)$ is a function that describes the relation between the states and the measurements;

$e(t)$ is measurement noise.

Given a discrete time t, we would like to get the estimator $\hat{x}(t|t-1)$ of the states $x(t)$ as good as possible using all available measurement data up to $t-1$.

A very common simplification is to assume that all functions used in the state space model are linear, and that both process noise and measurement noise are additive. This model looks like follows:

Measurement equation:

$$y_t = \underset{n_y \times 1}{Z_t} \underset{n_y \times n_x \, n_x \times 1}{x_t} + \underset{n_y \times 1}{e_t} ,$$ (9.29)

Transition equation:

$$x_t = \underset{n_x \times 1}{T_t} \underset{n_x \times n_x \, n_x \times 1}{x_{t-1}} + \underset{n_x \times 1}{w_t} ,$$ (9.30)

$$E\left(w_t w_\tau^T\right) = \begin{cases} Q_t & \text{for } t = \tau \\ 0 & \text{otherwise} \end{cases},$$ (9.31)

$$E\left(e_t e_\tau^T\right) = \begin{cases} R_t & \text{for } t = \tau \\ 0 & \text{otherwise} \end{cases},$$ (9.32)

$$E\left(w_t\right) = 0, \forall t ,$$ (9.33)

$$E\left(e_t\right) = 0, \forall t ,$$ (9.34)

where

x_t is the state vector and assumed to be generated by the first-order Markov process;

T_t is a transition matrix;

w_t is the process noise with mean zero and covariance matrix Q_t;

e_t is the measurement noise with mean zero and covariance matrix R_t.

The transition process is initiated with the random vector x_0, which has a mean of a_0 and a covariance matrix of P_0, that is $E[x_0] = a_0$ and $\text{Var}[x_0] = P_0$.

Thus Kalman Filter (described in detail in Chapter 1) is an efficient recursive algorithm for the computation of the optimal estimator of any unknown parameters in the linear state space system above on the basis of the observations $y_1, y_2, ..., y_T$.

9.4.1 Additive Non-Linear State Space model (Recap)

The specific nonlinear state space model we considered here is termed as the additive nonlinear state space model which can be represented as following:

Measurement equation:

$$y_t = h(x_t) + e_t \tag{9.35}$$

Transition equation:

$$x_t = f(x_{t-1}) + w_t, \tag{9.36}$$

$$E(w_t w_\tau^T) = \begin{cases} R_t & \text{for } t = \tau \\ 0 & \text{otherwise} \end{cases}, \tag{9.37}$$

$$E(e_t e_\tau^T) = \begin{cases} Q_t & \text{for } t = \tau \\ 0 & \text{otherwise} \end{cases}, \tag{9.38}$$

$$E(w_t e_\tau^T) = 0, \forall t, \tau, \tag{9.39}$$

where $f(\cdot)$ is the process model, x_t is the state of the system at time t and w_t is an additive process noise. It is assumed that the only information available about this system is a set of noisy observations y_t. These observations are related to the state vector by the nonlinear equation (9.35). $h(\cdot)$ is the observation model that transfers the state space vector into observation space, and e_t is an additive measurement noise. It is assumed that w_t and e_t are independent uncorrelated white noise.

The true states of the system are unknown and need to be estimated. We let $\hat{x}_{t+1|t}$ be the conditional mean at time $t+1$ conditional on all observations up to t that is

$$\hat{x}_{t+1|t} = E[x_{t+1} | y_t] \tag{9.40}$$

where $y_t = \{y_1, y_2, ..., y_t\}^T$. The estimated covariance is,

$$P_{t+1|t} = E\left[(x_{t+1} - \hat{x}_{t+1|t})(x_{t+1} - \hat{x}_{t+1|t})^T | y_t\right]. \tag{9.41}$$

The EKF (Extended Kalman Filter) is designed to estimate a subclass of nonlinear state-space models with additive, Gaussian process and measurement noises. The Extended Kalman Filter assumes that the errors in the state estimates are small. As a consequence, the predicted mean is approximated by,

$$\hat{x}_{t+1|t} = E\left[f\left(x_t\right)\middle|y_t\right]$$
$$\approx f\left[E\left(x_t\middle|y_t\right)\right] \qquad (9.42)$$
$$\approx f\left(\hat{x}_{t|t}\right).$$

That is, the predicted mean is equal to the prior mean projected through $f\left(\cdot\right)$. This estimate does not consider the actual distribution of the errors on the state prediction. Further, it is assumed that the state errors propagate through a separate linearized system and the covariance of these errors evolves according to,

$$P_{t+1|t} = \mathscr{J}_f P_{t|t} \mathscr{J}_f^T + Q_t \qquad (9.43)$$

where $\mathscr{J}_f$ is the Jacobian matrix of $f\left(\cdot\right)$ evaluated about $\hat{x}_{t|t}$.

Similar assumptions are made in predicting

$$\hat{y}_{t+1|t}, P_{yy,t+1|t}.$$

Unfortunately, the EKF suffers two well-known problems: first, the required Jacobian matrices, i.e., the matrix/linear approximations of nonlinear functions, can be extremely difficult and error-prone to derive. Second, the EKF linearized approximations can be extremely inaccurate and lead to filter instability.

9.4.2 The Scaled Unscented Transformation (Recap)

Unlike the EKF, the Unscented Kalman Filter (UKF) does not approximate the non-linear process. It uses the true nonlinear models and rather approximates the distribution of the state random variable. In the UKF the state distribution is still represented by a Gaussian random variable (GRV), but it is specified using a minimal set of deterministically chosen sample points. These sample points completely capture the true mean and covariance of the GRV, and when propagated through the true nonlinear system, captures the posterior mean and covariance accurately to the 2nd order for any nonlinearity, with errors only introduced in the 3rd and higher orders. In order to express UKF clearly, we have to introduce the following concepts.

The unscented transformation (UT) is a method for calculating the statistics of a random variable which undergoes a nonlinear transformation and builds on the principle that it is easier to approximate a probability distribution than an arbitrary nonlinear function (Julier and Uhlmann (1996)). Consider propagating a n_x dimensional random variable x through an arbitrary nonlinear function $f : \mathbb{R}^{n_x} \mapsto \mathbb{R}^{n_y}$ to generate y,

$$y = h(x) \tag{9.44}$$

Assume x has mean $\bar{x}$ and covariance P_x. To calculate the first two moments of y using the UT, we proceed as follows: first, a set of $2n_x + 1$ sigma points $\mathcal{S}_i = \{W_i, \mathcal{X}_i\}$ are deterministically chosen so that they completely capture the true mean and covariance of the prior random variable x. A selection scheme that satisfies this requirement is

$$\mathcal{X}_0 = \bar{x}, \ W_0 = \kappa / (n_x + \kappa), \ i = 0 \tag{9.45}$$

$$\mathcal{X}_i = \bar{x} + \left(\sqrt{(n_x + \kappa) P_x} \right)_i, \ W_i = 1 / \{2(n_x + \kappa)\}, \ i = 1, \ldots, n_x \tag{9.46}$$

$$\mathcal{X}_i = \bar{x} - \left(\sqrt{(n_x + \kappa) P_x} \right)_i, \ W_i = 1 / \{2(n_x + \kappa)\}, \ i = n_x + 1, \ldots, 2n_x \tag{9.47}$$

Each sigma point is now propagated through the nonlinear function,

$$\mathcal{Y}_i = h(\mathcal{X}_i), \ i = 0, \ldots, 2n_x \tag{9.48}$$

and the estimated mean and covariance of y are computed as follows,

$$\bar{y} = \sum_{i=0}^{2n_x} W_i \mathcal{Y}_i, \tag{9.49}$$

$$P_{yy} = \sum_{i=0}^{2n_x} W_i (\mathcal{Y}_i - \bar{y})(\mathcal{Y}_i - \bar{y})^T. \tag{9.50}$$

These estimates of the mean and covariance are accurate to the second order of the Taylor series expansion of $h(x)$ for any nonlinear function. Errors are introduced in the third and higher order moments but are scaled by the choice of the parameter κ. In comparison, the EKF only calculates the posterior mean and covariance accurately to the first order with all higher order moments truncated. The distance of the ith sigma point from $\bar{x}$, $|\mathcal{X}_i - \bar{x}|$ is proportional to $\sqrt{(n_x + \kappa)}$. For the special case of $\kappa = 3 - n_x$, the desired dimensional scaling invariance is achieved by canceling the effect of n_x. However, when $\kappa = 3 - n_x < 0$ the weight $W_0 < 0$ and the calculated covariance can be non-positive semi-definite. Then the scaled unscented transformation was developed to address this problem.

The scaled unscented transformation (SUT) replaces the original set of sigma points with a transformed set given by,

$$\mathcal{X}_i' = \mathcal{X}_0 + \alpha(\mathcal{X}_i - \mathcal{X}_0), \ i = 0,...,2n_x , \qquad (9.51)$$

where α is a positive scaling parameter which can be made arbitrarily small to minimize higher order effects.

A set of sigma points $\mathcal{S}_i = \{W_i, \mathcal{X}_i\}$ is calculated using equation (9.51) and then transformed into the scaled set $\mathcal{S}_i' = \{W_i', \mathcal{X}_i'\}$ by,

$$\mathcal{X}_i' = \mathcal{X}_0 + \alpha(\mathcal{X}_i - \mathcal{X}_0) \qquad (9.52)$$

$$W_i' = \begin{cases} W_0 / \alpha^2 + (1 - 1/\alpha^2), i = 0 \\ W_i / \alpha^2, i \neq 0 \end{cases} \qquad (9.53)$$

where α is the new sigma point scaling parameter. The process above can be combined into a single step by setting,

$$\lambda = \alpha^2 (n_x + \kappa) - n_x \qquad (9.54)$$

and selecting the sigma point set by:

$$\mathcal{X}_0' = \overline{x} \qquad (9.55)$$

$$\mathcal{X}_i' = \overline{x} + \left(\sqrt{(n_x + \lambda) P_x} \right)_i , \ i = 1,...,n_x \qquad (9.56)$$

$$\mathcal{X}_i' = \overline{x} - \left(\sqrt{(n_x + \lambda) P_x} \right)_i , \ i = n_x + 1,...,2n_x \qquad (9.57)$$

$$W_0^{(m)} = \lambda / (n_x + \lambda) \qquad (9.58)$$

$$W_0^{(c)} = \lambda / (n_x + \lambda) + (1 - \alpha^2 + \beta) \qquad (9.59)$$

$$W_i^{(m)} = W_i^{(c)} = 1 / \{2(n_x + \lambda)\}, \ i = 1,...,2n_x . \qquad (9.60)$$

The weighting on the 0th sigma point directly affects the magnitude of the errors in the forth and higher order terms for symmetric prior distributions (Julier (2000)). The introduced third parameter β affects the weighting of the 0th sigma point for the calculation of the covariance. This allows for the minimization of higher order errors if prior knowledge (i.e. kurtosis, etc.) of the distribution of x is available.

9.5 The Application of Unscented Kalman Filter

The Unscented Kalman Filter (UKF) is a straightforward application of the scaled unscented transformation to recursive minimum mean-square-error (RMMSE) estimation, (Julier and Uhlmann (1997b)). Following steps summarize the implementation process for the UKF.

Step 1: Selecting sigma points:
Generally, given a $n_x \times 1$ vector of latent states, a set of $2n_x + 1$ sigma points is selected, as

$$\mathcal{X}_{0,t-1} = x_{t-1|t-1} \qquad (9.61)$$

$$\mathcal{X}_{i,t-1} = x_{t-1|t-1} + \left(\sqrt{(n_x + \lambda)P_{t-1|t-1}}\right)_i, \, i = 1,...,n_x \qquad (9.62)$$

$$\mathcal{X}_{i,t-1} = x_{t-1|t-1} - \left(\sqrt{(n_x + \lambda)P_{t-1|t-1}}\right)_i, \, i = n_x + 1,...,2n_x \qquad (9.63)$$

where,

$$\lambda = \alpha^2 (n_x + \kappa) - n_x. \qquad (9.64)$$

α and κ are two user-specified constants. The constant α is a value between 0 and 1 which determines the spread of the sigma points away from $x_{t-1|t-1}$. If α is close to 0, then the sigma points will be very close to $x_{t-1|t-1}$, more of the higher-order nonlinearities are ignored. The constant κ is a positive constant used to scale the kurtosis of the sigma point distribution when desired. In practice, the value of κ is not critical and it is usually set to 0, or to $3 - n_x$ to ensure that the kurtosis of the sigma point distribution agrees with the kurtosis of a Gaussian distribution, (Julier and Uhlmann (2002); Wan and Van der Merwe (2001)).

Step 2: Get projected forward sigma points through dynamic transition equation:

After these sigma points have been selected, each of them is propagated forward in time through the dynamic equation as,

$$\mathcal{X}^*_{i,t|t-1} = f\left(\mathcal{X}_{i,t-1}\right), i = 0,1,...,2n_x . \qquad (9.65)$$

Step 3: Compute prior state and covariance estimates for latent state:

Prior state estimates $x_{t|t-1}$ and covariance estimates $P_{t|t-1}$ at time t are approximated by the weighted mean and variance of the transformed sigma points. The weights for the sigma points is defined as,

$$W_0^{(m)} = \lambda / \left(n_x + \lambda\right) \qquad (9.66)$$

$$W_0^{(c)} = \lambda / \left(n_x + \lambda\right) + \left(1 - \alpha^2 + \beta\right) \qquad (9.67)$$

$$W_i^{(m)} = W_i^{(c)} = 1 / \left\{2\left(n_x + \lambda\right)\right\}, i = 1,...,2n_x \qquad (9.68)$$

where β is another scaling constant explained above. For Gaussian distributions, $\beta = 2$ is optimal, Van der Merwe et al (2000). Then the prior state and covariance estimates are calculated using the sigma points and their associated weights as,

$$x_{t|t-1} = \sum_{i=0}^{2n_x} W_i^{(m)} \mathcal{X}^*_{i,t|t-1} , \qquad (9.69)$$

$$P_{t|t-1} = \sum_{i=0}^{2n_x} W_i^{(c)} \left[\left(\mathcal{X}^*_{i,t|t-1} - x_{t|t-1}\right)\left(\mathcal{X}^*_{i,t|t-1} - x_{t|t-1}\right)^T\right] + Q_t . \qquad (9.70)$$

Step 4: Predict observation:

Similar to the way we get forward sigma points, we get predicted observation by subjecting the sigma points to the measurement equation. More specifically, firstly augmenting the prior state sigma-point set to include the uncertainty constituted by the process noise as,

$$\mathcal{X}_{i,t|t-1} = \left[\mathcal{X}^*_{i,t|t-1}; \mathcal{X}^*_{i,t|t-1} + \lambda\sqrt{Q_t}; \mathcal{X}^*_{i,t|t-1} - \lambda\sqrt{Q_t}\right] \qquad (9.71)$$

and subsequently projecting the concatenated sigma-point set through the measurement equation as,

$$\mathcal{Y}_{i,t|t-1} = h\left(\mathcal{X}_{i,t|t-1}\right). \qquad (9.72)$$

Predicted observations and the associated variance and covariance matrices are computed as,

$$y_{t|t-1} = \sum_{i=0}^{2n_x} W_i^{(m)} \mathcal{Y}_{i,t|t-1} \tag{9.73}$$

$$P_{y_t y_t} = \sum_{i=0}^{2n_x} W_i^{(c)} \left[\left(\mathcal{Y}_{i,t|t-1} - y_{t|t-1} \right) \left(\mathcal{Y}_{i,t|t-1} - y_{t|t-1} \right)^T \right] + R_t \tag{9.74}$$

$$P_{x_t,y_t} = \sum_{i=0}^{2n_x} W_i^{(c)} \left[\left(\mathcal{X}_{i,t|t-1} - x_{t|t-1} \right) \left(\mathcal{Y}_{i,t|t-1} - y_{t|t-1} \right)^T \right]. \tag{9.75}$$

Step 5: Kalman Filter recursion:

The discrepancy between model prediction and actual observations is weighted by a Kalman gain function to yield posterior state and covariance estimates,

$$K_t = P_{x_t,y_t} P_{y_t y_t}^{-1} \tag{9.76}$$

$$x_{t|t} = x_{t|t-1} + K_t \left(y_t - y_{t|t-1} \right) \tag{9.77}$$

$$P_{t|t} = P_{t|t-1} - K_t P_{y_t y_t} K_t^T. \tag{9.78}$$

The gain and update equations are formulated in equivalent form to equations in the linear case, except that they are composed of elements computed using the aforementioned weighted sigma points. Step 1 to step 5 are then repeated until all the observations have been incorporated.

9.6 Empirical Results

This part provides the empirical results for our state-space model calibration, which is important for us to understand the characteristic of implied volatility in Markit iTraxx market and its relationship with realized volatility.

Table 9.1 lists the estimated results using the entire sample data from Oct 23, 2006 to Apr 14, 2008. Table 9.2 lists the estimated results using our two separated sub-sample periods. Firstly, we examine the Table 9.1. As mentioned above that parameter ρ can be interpreted as the correlation between the log-return and volatility of the asset which affects the heaviness of the tail. For most empirical results from stock market, ρ is usually negative and that means when return goes up, the volatility goes down. But for CDS market, this relationship usually reverse because increasing CDS level or changes indicate more uncertainty in the future market, then the volatility usually goes up. This is consistent with the general finding that the volatility in downtrend market is much higher than the volatility in uptrend market. From the estimators in Table 9.1, we can see that most of estimators of ρ are positive and very close to 1. The exceptions are for the implied volatility of iTraxx Europe Crossover (3-month and 9-month maturities) and the implied volatility of iTraxx Europe (6-month maturity). The exceptions for iTraxx Europe Crossover may not be so surprising. This index includes 50 most liquid sub-investment entities and it is quite different from the other two benchmark indices.

When we move to estimators of ρ in Table 9.2, there are two interesting points we want to address here. First, the estimators of ρ are very different for two different sample periods. During the first half period of our sample, the correlation between the daily logarithmic change of iTraxx Europe and its volatility is negative. Similarly, the correlation between the daily logarithmic change of iTraxx Europe Hivol and its volatility is also negative in the same period.

But these correlations become positive during the second half period of our sample. During this time, the whole CDS market endured the credit crunch resulting from the sub-prime mortgage collapse in the U.S. Another interesting point is that the correlation between the daily logarithmic change of iTraxx Europe Crossover and its volatility is always negative corresponding to 3-month and 9-month implied volatilities. But this correlation is positive for the 6-implied volatility.

Based on above analysis the basic conclusion we can make here is that the correlations between the daily logarithmic change of investment grade CDS rates and their volatilities are positive. This is especially so when the market indicates a perceived deterioration of a company's creditworthiness. This may be due to the fact that during the normal or up-trending market, increased CDS rate for investment grade companies would not be of much concern for the market. The market would expect more potential of revival of these companies rather than possibility of further deterioration. This kind of perception reverses when the whole market is in down-trending mode.

But the correlation between the daily logarithmic change of sub-investment grade CDS and their volatility are negative which may reflect the fact that the market perceives higher probability of rebound than default, at least for those relatively more liquid sub-investment grade entities.

Parameter $\bar{v}^*$ can be interpreted as the long-run variance, consistent to the initial analysis for iTraxx CDS indices. The long-run variance of logarithmic daily changes of iTraxx Europe is larger than that of iTraxx Europe Crossover. Also, this kind of variance is usually larger for shorter term maturity option than longer term maturity option. This may indicate that the time series of iTraxx CDS indices are mean-reverting.

For example, the estimators of $\bar{v}^*$ from the entire sample are 0.9686, 0.9501 and 0.3426 for 3-month, 6-month and 9-month iTraxx Europe 5-year options. At the same time these numbers become to 0.6773, 0.4942 and 0.207 respectively for the underlying iTraxx Europe Crossover 5-year index. Comparing the estimators in the two panels of Table 9.2, we notice that these estimators are usually larger for the second sub-period of sample. That makes sense due to the onset of credit crunch.

Parameter σ_v affects the kurtosis of the distribution, which is also the volatility of volatility of the underlying asset. For simplicity, let, V_M, V_H and V_C denote the variance of daily forward logarithmic change of iTraxx Europe, iTraxx Europe Hivol and iTraxx Europe Crossover respectively. All these iTraxx CDS indices are traded with 5 years maturity. Then $\sigma_{v,M}$, $\sigma_{v,H}$ and $\sigma_{v,C}$ denote the volatility of V_M, V_H and V_C respectively. When comparing σ_v across three different iTraxx CDS indices for the entire sample period as well as for each of the two different sub-sample periods, we can clearly see that $\sigma_{v,C}$ is the smallest. The difference between $\sigma_{v,H}$ and $\sigma_{v,M}$ seems trivial from the results based on two separated sub-samples. But $\sigma_{v,H}$ seems a little bit higher than $\sigma_{v,M}$ for the whole sample period.

This characteristic of volatilities of volatilities can be verified using historical Markit iTraxx CDS indices data. The difference between these volatilities of volatilities is more obvious during the first half period of our sample than that during the second half period. When market is turbulent, all these volatilities of volatilities are quite high and their difference is trivial.

These findings above indicate economy changes, at least less favorable economy changes, impact higher credit rating firms more seriously. When we compare σ_v across different maturities, we find that usually the implied volatilities of volatilities with shorter maturity is higher than its longer maturity counterpart. This is mainly because we assume that the variance is a mean-reverting process. This mean-reversion process assumption is quite reasonable if we examine the characteristic of realized variance of logarithmic changes of iTraxx indices. Table 9.3 demonstrates that the standard deviation of 3-month realized variance of iTraxx Europe Crossover daily changes is 19.25% while it is 37% for that of iTraxx Europe daily changes. Comparing this statistic across different measuring period, the standard derivation of short-term variance is larger.

Parameter κ_v^* is the speed of mean reversion parameter. Generally, this parameter is larger for short-term variance. For example, for the entire sample, the κ_v^* for 3-month implied volatility of iTraxx Europe 5 years is 1.894, while it is 0.9914 for 9-month implied volatility of iTraxx Europe 5 years. This means that if the variance of logarithmic daily iTraxx indices changes away from its long-term mean, then it revert to its long-run level more quickly for shorter term variance. Or in other words, the longer term variance is more clustering.

Figure 9.1 plots the estimated time series of implied volatility and Figure 9.2 plots the realized volatility for comparison. Table 9.4 lists the summary statistics for these implied volatilities and realized volatilities. Obviously, there is a close link between the realized volatility and the implied volatility. It is not surprising, that the implied volatilities in Figure 9.1 demonstrate two different patterns. The implied volatilities after July 2007 are quite different from that of 6 months before. Roughly speaking, before July 2007 the whole CDS market is quite stable, the average implied volatility is only 14%-20% which is lower than contemporaneous realized volatility. The implied volatility starts to surge and fluctuate more widely since July, 2007. The average implied volatility is larger than contemporaneous realized volatility during the time period between July 13, 2007 and Apr 14, 2008, although it goes below the contemporaneous realized volatility at times.

9.7 Concluding Remarks

The objective of this chapter has been to demonstrate the application of stochastic filtering methodology using the unscented Kalman Filter. This has been carried out in the context of option pricing model for the underlying CDS. The option pricing model assumes the presence of stochastic volatility as opposed to constant volatility as in the case of Black-Scholes model for stock options.

The option pricing model is calibrated to observed market prices and thus allowing us to infer the implied volatility parameters. This technique makes it possible to estimate time series of implied volatility from market prices. By comparing the estimated implied volatility parameters and the realized volatilities, we conclude that our estimation is reasonable.

Table 9.1 Estimated Results Form State Space Model Described in Section 9.3.

Underlying iTraxx	Option Maturity	κ_v^*	$\overline{v}^*$	ρ	σ_v	Errors	Log Likelihood
Europe 5 Year	3 Months	1.8941	0.9686	0.9998	0.7698	2.23E-05	1366.98
	6 Months	0.5620	0.9501	-0.8608	0.9974	2.72E-08	1517.29
	9 Months	0.9914	0.3426	0.9997	0.6310	6.68E-08	1516.79
Hivol 5 Year	3 Months	0.9648	0.8714	0.9811	0.8281	2.30E-09	1491.10
	6 Months	0.7059	0.7916	0.9999	0.9377	9.72E-07	1326.89
	9 Months	0.3750	0.9668	0.9995	0.7848	3.73E-09	1403.77
Crossover 5 Year	3 Months	1.0403	0.6773	-0.5382	0.6839	2.06E-09	1005.52
	6 Months	0.6800	0.4942	0.9996	0.8186	9.60E-07	749.70
	9 Months	1.1714	0.2017	-0.1853	0.5216	2.94E-07	857.83

The data used for estimation covers period form Oct 23, 2006 to Apr 14, 2007. The estimated variances are quite small for all these estimators, and are approximately around 1.0E-5.

Table 9.2 Estimated Results Form State Space Model Described in Section 9.3.

Underlying iTraxx	Option Maturity	κ_v^*	$\overline{v}^*$	ρ	σ_v	Errors	Log Likelihood
Panel A							
Europe	3 Months	2.6862	0.3419	-0.9864	1.0000	5.65E-08	-1000.15
5 Year	6 Months	0.4705	0.9431	-0.9988	0.8088	7.70E-09	-869.75
	9 Months	1.4258	0.3043	-0.9246	0.7657	2.11E-09	-965.29
Hivol	3 Months	2.7101	0.3940	0.6853	0.7177	2.33E-08	-896.70
5 Year	6 Months	1.6128	0.3239	-0.9764	0.9883	5.06E-06	-725.51
	9 Months	0.6558	0.6566	-0.9664	0.6484	1.35E-07	-846.05
Crossover	3 Months	0.5565	1.0000	-0.9999	0.4998	2.06E-09	-592.05
5 Year	6 Months	0.7482	0.5769	0.8156	0.8505	2.45E-05	-423.76
	9 Months	0.4916	0.5799	-0.8513	0.4862	1.79E-06	-537.65
Panel B							
Europe	3 Months	0.8046	0.6211	0.9975	0.9990	9.72E-09	-678.90
5 Year	6 Months	0.5461	0.9882	0.9298	0.9994	5.64E-07	-623.79
	9 Months	0.4221	0.9903	0.9999	0.6661	2.42E-07	-589.68
Hivol	3 Months	2.1450	1.0000	-0.9999	0.9997	1.44E-05	-624.15
5 Year	6 Months	0.9617	0.6802	0.9998	1.0000	2.14E-06	-549.03
	9 Months	0.4025	0.6291	0.9994	0.7082	2.80E-09	-571.71
Crossover	3 Months	2.7011	0.9953	-0.9975	0.9733	3.54E-09	-413.51
5 Year	6 Months	0.8764	0.1807	0.9999	0.5527	4.95E-09	-368.98
	9 Months	1.5327	0.8041	-0.9957	0.9776	4.53E-06	-349.96

Panel A gives results for data form October 23, 2006 to July 13, 2007; Panel B gives results based on data from July 16, 2007 to April 14, 2007.

Table 9.3 Descriptive Statistics of Realized Volatility.

iTraxx Index	3-Month	6-Month	9-Month
Europe	0.3700	0.3256	0.2901
Hivol	0.2871	0.2447	0.2197
Crossover	0.1925	0.1417	0.1231

This table reports the standard deviation of realized variance of daily logarithmic changes for three different iTraxx indices. The number in column '3-Month' reports the standard deviation of 3-Month realized variance (annualized) of iTraxx indices, so for column '6-Month' and '9-Month'.

Table 9.4 Descriptive Statistics of Volatility.

iTraxx Index	Expiry	Mean	St. Dev	Kurtosis	Skewness	Minimum	Maximum
Panel A (Oct 23, 2006-Apr 14, 2008)							
Implied Volatility							
	3M	70.7690	63.9314	-1.0364	0.4903	1.8296	231.924
Europe	6M	57.6652	44.1955	-0.7673	0.4932	8.7095	193.074
	9M	50.1249	38.7566	-1.7361	0.2516	10.467	115.562
	3M	60.5631	51.5784	-0.5789	0.5140	1.4065	245.226
Hivol	6M	49.8581	33.1507	-0.8149	0.4263	11.154	150.713
	9M	45.5202	30.3169	-1.7029	0.2111	12.198	96.2921
	3M	38.2051	23.0215	-0.8634	0.3532	7.9151	100.434
Crossover	6M	32.9807	20.5646	-1.0269	0.1834	4.0679	88.943
	9M	34.2522	18.9227	-1.4034	0.1444	8.6629	76.694
Realized Volatility							
	3M	62.3917	36.9886	-0.9508	0.6178	18.6833	135.532
Europe	6M	61.1675	32.5573	-1.7754	0.1849	23.9137	107.550
	9M	58.3899	29.0059	-1.7840	0.2087	27.6791	105.662
	3M	58.4555	28.7113	-0.9642	0.5496	19.4233	114.833
Hivol	6M	56.9471	24.4686	-1.7984	0.1332	27.0745	91.282
	9M	54.4283	21.9663	-1.6986	0.2524	29.4655	93.944
	3M	50.3423	19.2489	-0.7861	0.1606	18.8740	88.830
Crossover	6M	51.1627	14.1744	-1.2567	-0.1302	27.6236	71.418
	9M	50.3872	12.3094	-1.6190	-0.0492	30.8010	66.984

Table 9.4 Descriptive Statistics of Volatility (Contd.).

iTraxx Index	Expiry	Mean	St. Dev	Kurtosis	Skewness	Minimum	Maximum
Panel B (Oct 23, 2006-July 13, 2007)							
Implied Volatility							
	3M	12.1523	7.3958	-0.9894	0.4290	1.8296	34.116
Europe	6M	17.8404	10.8930	4.3523	2.1233	8.7095	58.347
	9M	13.7756	2.8905	-1.5325	0.2494	10.4671	21.072
	3M	13.6605	11.5633	0.6860	0.9756	1.4065	60.100
Hivol	6M	19.7879	9.1515	2.1527	1.5660	11.1547	50.820
	9M	16.6794	3.8394	-1.4873	0.2835	12.1981	26.006
	3M	17.7420	8.2350	-0.7502	0.5764	7.9151	42.689
Crossover	6M	16.0149	13.3663	7.3120	2.4472	4.0679	76.718
	9M	16.6437	6.3120	0.0485	0.8111	8.6629	37.391
Realized Volatility							
	3M	29.9738	5.7130	-0.8552	0.0927	18.6833	44.565
Europe	6M	30.0152	3.4760	-0.4545	-0.1918	23.9137	40.758
	9M	30.7769	1.4034	1.6727	0.5481	27.6791	36.996
	3M	33.9806	7.8993	-1.1948	0.1078	19.4233	45.920
Hivol	6M	33.4351	3.6950	-1.0537	-0.4100	27.0745	42.000
	9M	33.7123	1.3718	1.2944	-0.9454	29.4655	37.786
	3M	39.6345	16.3080	-1.4062	0.2836	18.8740	62.852
Crossover	6M	38.9976	7.8695	-1.5768	-0.1114	27.6236	53.154
	9M	38.7318	4.2626	-1.3182	-0.4844	30.8010	44.995

Table 9.4 Descriptive Statistics of Volatility (Contd.).

iTraxx Index	Expiry	Mean	St. Dev	Kurtosis	Skewness	Minimum	Maximum
Panel C (July 16, 2007-April 14, 2008)							
Implied Volatility							
	3M	127.757	37.6758	0.6483	0.5507	39.1332	231.924
Europe	6M	96.3291	26.5080	1.9468	0.9159	46.0072	193.074
	9M	85.5011	20.3909	2.2988	-1.4310	22.4433	115.562
	3M	106.008	30.2852	3.2757	1.3389	47.6362	245.226
Hivol	6M	79.0490	18.8757	2.5774	1.0574	42.5935	150.713
	9M	73.5691	14.2147	1.6895	-1.0466	27.7365	96.292
	3M	58.0064	13.4564	0.8221	0.8288	36.2990	100.434
Crossover	6M	49.3989	10.6047	1.6509	0.9478	31.7576	88.943
	9M	51.2976	8.7147	-0.2314	0.3464	36.9601	76.694
Realized Volatility							
	3M	94.1122	25.2595	-1.2374	0.3488	45.4481	135.532
Europe	6M	82.4211	20.2065	-1.2250	0.1629	40.9680	114.833
	9M	60.8466	15.9138	-1.2641	0.3603	38.4667	88.830
	3M	91.6058	14.4612	2.3181	-1.3820	41.4397	107.550
Hivol	6M	79.9182	10.1309	3.4991	-1.6776	42.1096	91.282
	9M	63.0128	7.2403	-1.1570	-0.6909	48.1638	71.418
	3M	85.4086	13.5884	3.3561	-1.5693	37.4710	105.662
Crossover	6M	74.7067	10.9696	2.4082	-1.1959	37.9081	93.944
	9M	61.7715	4.4300	2.8160	-1.5239	45.0794	66.984

Table 9.4 reports basic descriptive statistics for the volatility data. Realized volatility is computed based on three different spot 5 years iTraxx indices. All these volatilities are in annualized standard deviation in percentage units. The three panels (A, B and C) relate to different sample periods as indicated.

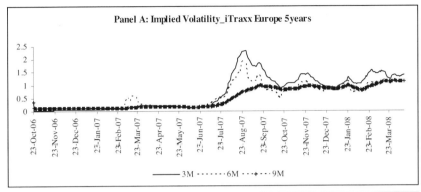

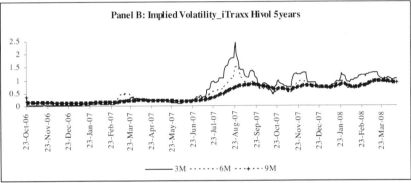

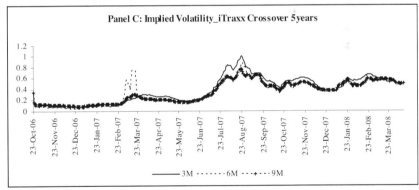

Figure 9.1 Time-Series Plot of Implied Volatility.

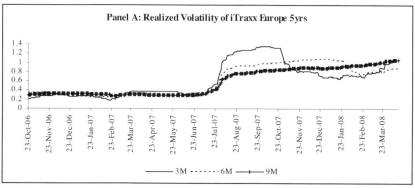

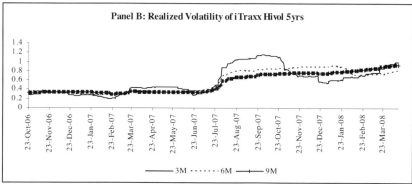

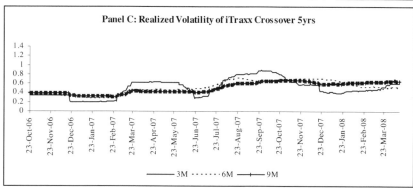

Figure 9.2 Time-Series Plot of Realized Volatility.

Chapter 10

Stochastic Volatility Model and Non-Linear Filtering Application

This chapter focuses on advanced estimation procedures for models that require stochastic filtering. This is necessitated by the fact that these models do not satisfy standard assumptions of linear Kalman filtering. The classic examples are related to various forms of stochastic volatility models reported in the literature. Three specifications for stochastic volatility models of interest rates are considered in this chapter - AR(1), ARMA(1,1) and AR(MA)X specifications. These specifications of the models are estimated using different approaches e.g. Quasi-Maximum Likelihood (QML) and Monte Carlo Likelihood (MCL) function. We also relate this to the application of Particle Filter (PF) already discussed in a related publication by the author. Relative performances of these approaches are also discussed.

Most reported stochastic volatility (SV) model for interest rates only deal with an AR specification for the latest factor process. We show in this chapter how to accommodate alternative time series dynamics in the latent process. We demonstrate the efficacy of this approach with an application to U.S. short-term interest rate data using two different estimation techniques e.g. QML (Quasi Maximum Likelihood) and MCL (Monte Carlo Likelihood). We find that the elasticity parameter of the variance is closer to 0.5 i.e. similar to that in Cox-Ingersoll-Ross (1985) model of interest rates. This is quite a contrast to the finding in Chan, Karolyi, Longstaff, and Sanders (1992).

10.1 Background to Stochastic Volatility Models

Modeling the term structure of interest rates is a critical component in asset pricing and has been an important topic for many years. It has been well established in the literature that interest rates, at least in the U.S., exhibits time-varying volatility where the volatility itself has been found to be volatile. Ball and Torous (1999), Smith (2002) and Sun (2005) illustrate the stochastic nature of interest rate volatility. They incorporate stochastic volatility to short term interest rate models based on the discrete-time short rate models of Chan et al. (1992, hereafter CKLS). The studies found significant persistence in volatility although the mean-reversion was generally found to be faster than that of stocks.

Stochastic volatility (SV) models in discrete time have been traditionally applied to stock market returns as an alternative to GARCH models, though they remain largely less popular due to the difficulty in evaluating the likelihood function. In these models, volatility is defined as a random, unobserved latent variable of an underlying return series, where the log of the squared residuals is most commonly modeled as a first-order autoregression. However, exact evaluation of the likelihood function is not possible resulting in a number of techniques to estimate it.

The earliest techniques include the quasi-maximum likelihood (QML) method introduced by Harvey et al. (1994) where the assumption of a normal distribution is made when the distribution is non-normal. This less efficient method of model estimation has its advantages, the primary being the formulation of the model in state space form which allows relatively straightforward algorithms for filtering and smoothing to be applied. Other more accurate techniques have been proposed including the Jacquier et al. (1994) Monte Carlo Markov Chain (MCMC) technique or the Monte Carlo Likelihood (MCL) of Sandmann and Koopman (1998). In the previous studies of short rate volatility, the methods employed include the use of QML (Smith (2002) and Sun (2005)), a non-Gaussian estimator (Ball and Torous (1999)) and MCMC (Kalimipalli (2004)) in the estimation of short rate volatility.

This analysis implements the MCL method of estimation as it not only uses QML estimation as an intermediate step, it is computationally efficient compared to the other existing methods of SV likelihood estimation. As a consequence, this sudy makes a comparison of the QML and MCL estimation techniques for SV model estimation.

In general, the traditional SV model specifies the log-variance as an AR(1) process. It has been suggested that the model can also be specified as higher order ARMA models though few studies have investigated this claim. The SV-ARMA(1,1) model is introduced in this study as an extension to the AR(1) specification. It is determined whether the ARMA(1,1) specification provides a better estimation of stochastic volatility over a traditional AR(1) specification of the volatility for a short-term interest rate process.

In addition, this analysis demonstrates incorporating exogenous variables in the volatility equation in the empirical study. It considers the term spread and meetings of the Federal Open Market Committee (FOMC) as factors that not only influence the level of short-term interest rates but also the volatility. Whilst both factors proved to be statistically significant, the term spread only had a marginal effect on the volatility. However, the FOMC meetings provided a much stronger impact where it was found that FOMC meetings tended to increase the volatility exhibited in the yields of short term bonds.

A key finding of this study is that contrary to CKLS and Smith (2002), the nested model was found to support the Cox et al. (1985, hereafter CIR) short term interest rate process. If the short rate process follows the CIR model, then the elasticity of variance parameter should be around 0.5 as opposed to the value of close to 1.5 reported in both CKLS and Smith (2002) even though the latter incorporates a stochastic volatility model. This finding is more consistent with Ball and Torous (1999) and Sun (2005) where they found the elasticity was around 0.65 for U.S. treasury bonds.

10.2 Stochastic Volatility Models of Short-term Interest Rates

Following the same methodology as Ball and Torous (1999) and subsequently Smith (2002) and Sun (2005), the short rate is modeled according to the CKLS discretization of the stochastic differential equation:

$$dr_t = (a + br_t) + \sigma_t r_t^\gamma dZ_t \qquad (10.1)$$

where Z is Brownian motion, by

$$\Delta r_t = a + br_{t-1} + \sigma_t r_{t-1}^\gamma \varepsilon_t \qquad (10.2)$$

where $\varepsilon_t \sim NID(0,1)$ and γ is generally regarded as the volatility elasticity parameter. The stochastic differential equation in (10.1) nests a number of continuous time short-term interest rate models and provided a framework for comparing these models. In CKLS, the volatility parameter σ_t was assumed constant and γ was set at a specified value to match the values provided by the various short rate models. These assumptions have been relaxed in this analysis. In particular, the volatility process itself is driven by a stochastic process and is modeled according to Ball and Torous (1999) and Brenner, Harjes and Kroner (1996) based on the specification of Taylor (1986). This specification allows the log-variance to evolve as a first-order autoregressive process, and by setting $e_t = \sigma_t r_t^\gamma \varepsilon_t$, this gives,

$$\log \sigma_t^2 \equiv h_t = \mu(1 - \varphi) + \varphi h_{t-1} + \eta_t \qquad (10.3)$$

where $\eta_t \sim NID(0, \sigma_\eta^2)$. It can be seen from the above equation that as $\log \sigma_t^2$ contains the error term η_t, it evolves stochastically and reverts to a long-term mean where the persistence in volatility is governed by φ. A drawback of squaring the residuals is that the information required in determining any correlation between the two errors ε_t and η_t is lost. However, it was found in CKLS and Sun (2005) that the "leverage effect" evident in stock returns (i.e. volatility of returns increases when returns are negative) is not evident in interest rates. Hence, by assuming zero correlation in the error terms, no loss of efficiency should be exhibited.

As the model can be cast into state space form, it allows for estimation using techniques available in the literature. A review of the main estimation techniques for stochastic volatility models can be found in Broto and Ruiz (2004). The method employed by a number of empirical studies on stochastic volatility and in particular, Smith (2002) and Sun (2005) use QML in their studies of short term interest rate volatility. QML is the most computationally efficient however it assumes that the model is a linear Gaussian state space model and applies the

Kalman filter to estimate the likelihood function. Though this estimation procedure is not efficient, Ruiz (1994) shows good finite sample properties of QML estimation and it has been argued that it is adequate in empirical studies. However, numerous studies have indicated that the fitted values found under QML estimation result in a time series of volatility that is too smooth and hence performs poorly in forecasting compared to other estimation techniques.

As an alternative, this study uses the MCL technique to estimate the models as it is relatively computationally efficient compared to the other non-Gaussian estimation techniques such as MCMC and has been shown to be more efficient than QML as it provides a better estimation of the likelihood function. The main technique employed by the MCL methodology is importance sampling via a simulation smoother. To compute the likelihood function, firstly the linear Kalman filter and smoother is applied using an importance density (usually a Gaussian density) that is close to the true distribution. The likelihood function is estimated as the likelihood of the importance density plus the ratio of the difference between the true density and the importance density using Monte Carlo simulation. An outline for MCL is given in a later section.

10.2.1 SV-ARMA Specification

As an extension to the traditional AR(1), we propose that the log-volatility can evolve as a more general autoregressive moving average process. For an ARMA(1,1) process, the log-variance process becomes:

$$\log \sigma_t^2 \equiv h_t = \mu(1-\varphi) + \varphi h_{t-1} + \theta \eta_{t-1} + \eta_t . \qquad (10.4)$$

By letting $e_t = \Delta r_t - (a + b r_{t-1})$ it can be shown that the volatility of the short rate with an ARMA(1,1) volatility process can be specified in state space form as follows:

$$\log e_t^2 = \begin{bmatrix} 1 & 0 \end{bmatrix} \begin{bmatrix} h_t \\ \theta \eta_t \end{bmatrix} + 2\gamma \log r_{t-1} + \log \varepsilon_t^2 , \qquad (10.5)$$

$$\begin{bmatrix} h_t \\ \theta \eta_t \end{bmatrix} = \begin{bmatrix} \mu \\ 0 \end{bmatrix}(1-\varphi) + \begin{bmatrix} \varphi & 1 \\ 0 & 0 \end{bmatrix} \begin{bmatrix} h_{t-1} \\ \theta \eta_{t-1} \end{bmatrix} + \begin{bmatrix} 1 \\ \theta \end{bmatrix} \eta_t . \qquad (10.6)$$

Hence the estimation techniques available such as QML and MCL are applicable as both techniques exploit the fact that the system can be specified in a state space form.

10.2.2 Exogenous Variables

In addition, if additional exogenous variables are thought to affect the volatility, then these can be included. For instance, Christiansen (2005) employed a bi-variate GARCH model to demonstrate that the volatility of the yield term spread (i.e. the difference between the yield of a short term bond and a long term bond) the volatility of short-term interest rates were correlated. Taking this notion further, the log-variance of the yield term spread as an exogenous variable in the log-variance equation of the short rate. The term spread can be included in the dynamics of the log-variance as follows:

$$h_t = \mu(1-\varphi) + \varphi h_{t-1} + \delta(1-\varphi L)s_{t-1} + \eta_t, \qquad (10.7)$$

$$h_t = \mu(1-\varphi) + \varphi h_{t-1} + \delta(1-\varphi L)s_{t-1} + \eta_t + \theta\eta_{t-1}, \qquad (10.8)$$

where s_t is the yield term spread at time t defined above, δ_s captures its impact on the volatility of the short-term interest rate process and L is the lag operator.

Alternatively, some studies have considered the impact of US Federal Reserve announcements on short-term interest rates as the yields on short-term bonds tend to closely track the Fed Funds rate. Das (2002) considers the effect of the Federal Open Market Committee (FOMC) meetings on the jump probability of short rates where an increase in jump probability corresponds to an increase in the volatility of the short rate. The findings suggest that the effect of 2-day FOMC meetings increased the probability of jumps occurring. Using a similar approach, we incorporate a dummy variable for the FOMC meetings to determine whether there is an impact on volatility. In this case, the value is 1 if there is an FOMC meeting in the preceding period and zero otherwise. This is incorporated in the SV models as follows,

$$h_t = \mu(1-\varphi) + \varphi h_{t-1} + \kappa f_{t-1} + \eta_t, \qquad (10.9)$$

$$h_t = \mu(1 - \varphi) + \varphi h_{t-1} + \kappa f_{t-1} + \eta_t + \theta \eta_{t-1}. \qquad (10.10)$$

In both of the above situations, estimation is done through maximum likelihood as before by simultaneously maximizing with respect to δ or κ and the parameters of the stochastic volatility model.

10.3 Data for Analysis

The data for the empirical analysis is obtained from DataStream. The data set consists of weekly observations of three-month constant maturity U.S. treasury bills from January 1990 to December 2006 giving a sample size of 856 observations. Additionally, the yield spread is calculated as the difference between the yield on a ten-year constant maturity U.S. Treasury bill (FRTCM10) and the yield on a three-month bill over the same period. The FOMC meeting dates are obtained from the St. Louis Federal Reserve website.

The summary statistics for the 3-month U.S. Treasury bill and residuals of the CKLS discretization can be found in Table 10.1 and plots of the same series can be found in Figures 10.1(a) and (b). On inspection of the plots of the series, the yields exhibit periods of both steep increases and steep decreases whilst also having periods when yields have remained relatively flat. These characteristics provide a good indication that not only is the volatility of the short rate time-varying, but it also exhibits some volatility clustering.

A similar conclusion can be gained from Figure 10.1(b), where a high volatility period is followed by a number of high volatility periods. Two such periods of high volatility and steep decreases in yields can be identified as during the Iraq war around 1990-92 and after September 11, 2001. Using more formal test statistics on the data, it can be seen from the augmented Dickey-Fuller unit root test statistic that it cannot be rejected at the 5% level that the yields of short-term bonds are an integrated process, whilst the residuals of the CKLS model are stationary. The short rate process is non-normal as indicated by the Jarque-Bera statistic, however the residuals of the CKLS model exhibit high kurtosis and the Jarque-Bera statistic indicates extreme non-normality. There is also strong evidence of autocorrelation and autoregressive conditional heteroscedasticity for both the short rate and the residuals. Looking at the estimated coefficients of the CKLS discretized model in Table 10.2, as expected a negative value of b was found implying that the short rate process is mean-reverting, though this is much lower than the estimate found in CKLS. However, this value is similar to other studies on stochastic volatility models of short-term interest rates.

10.4 Analysis of Estimation Results

The results of the maximum likelihood parameter estimates for the SV models can be found in Table 10.3 and Table 10.4. The estimates for both the QML and MCL technique are included allowing for some comparison between previous studies that only implement QML in estimating interest rate volatility.

When comparing the results of the two estimation procedures, it can be seen there are large differences in the parameter estimates. As σ_η^2 is the main driver in the stochastic nature of stochastic volatility models, the differences in the values of σ_η^2 can lead to significant differences in the conditional variances estimated by the two methods. This has been discussed in Ruiz (1994) and Broto and Ruiz (2004) where it is suggested that the lower the true value of σ_η^2, the worse QML estimation performs. Although this sample size is not small with around 855 data points and the σ_η^2 estimated under QML was found to be around 0.8-0.9, the MCL estimates were much lower with values around 0.6 and under. This indicates that whilst the results of QML estimation are indicative of the main features of the data set, the resultant parameter estimates are not entirely reliable. An example of the differences evident as a result of using QML over more accurate procedures is the case found in Smith (2002). That study uses monthly observations of 30-day treasury bills where it was found that many of the parameter estimates reported were not found to be statistically significant.

Concentrating on the value of the estimates themselves, it can be seen from all estimates of the autoregressive parameter that there is some persistence in volatility exhibited, with all values lying around 0.6 under

QML and between 0.6-0.7 for MCL. This is lower than the persistence found in stock returns indicating that interest rates have a faster mean-reversion than stocks where the persistence parameter has been found to be very high. This implies that shocks affecting the volatility of interest rates will not persist for as long a period of time and that volatility will generally revert to its long-term mean fairly quickly. The values for the volatility of volatility parameter, σ_η, was found to have values around 0.85-0.9 for the AR models and 0.7-0.75 for the ARMA models under QML estimation. Under MCL estimation, the AR models had values of σ_η around 0.65-0.7 and the ARMA models had values around 0.5. This indicates that when an MA term is introduced in the volatility equation, the volatility of volatility tends to be reduced. Hence some of the variation exhibited by the volatility can be explained by the correlation in the errors of the volatility component.

We now focus on the estimates of the elasticity parameter. From the specification of the model, it can be seen that the conditional variance of the interest rate process is proportional to $r_t^{2\gamma}$. The estimate of all models shows that the elasticity parameter lying around 0.65 for QML and 0.59 for MCL. This implies that the unexpected change in the interest rate process is proportional to about 1.2 the power of the level of interest rates. This is consistent with the studies by Ball and Torous (1998) and Sun (2005) that find the elasticity parameter to be also around 0.65 for U.S. interest rates. The CKLS estimate of $\gamma = 1.5$ is suggested by Duffee (1993) as being too high and the more recent estimate of $\gamma = 1$ by Smith (2002) that also incorporates stochastic volatility is also fairly high. The parameter estimated in this study is much closer to the original CIR specification where $\gamma = 0.5$ indicating that the variance is proportional to the level of interest rates. It can therefore be seen that both formulations of the stochastic volatility process allow credible estimation of this crucial parameter of the interest rate model.

When including the term spread as an exogenous variable in the volatility equation, the QML estimates of this parameter is only significant in the AR specification whereas the MCL estimates are both significant at the 5% level. As to whether the term spread is a driving factor of interest rate volatility, its impact appears to be marginal as the coefficient has a value of around 0.1 under MCL and 0.03-0.04 under

QML. The relationship between volatility and FOMC meetings is much stronger where a finding similar to Das (2002) was found as FOMC meetings tend to increase the volatility of short term interest rates. The coefficient was found to be significantly positive under both QML and MCL estimation with a value of around 0.42 found under QML and about 0.35 under MCL for both AR and ARMA specifications. This indicates that FOMC meetings provide market participants with additional information where a possible cause of the increase in volatility is due to an increase in trading activity around the time of each FOMC meeting.

10.5 Comparison of Volatility Estimates

Turning now to the performance of the models, we consider a number of different tests. We start with a comparison of the Akaike Information Criterion (AIC) and the Schwartz Bayesian Information Criterion (BIC) values for the competing models where the lower value implies a better the model. Using both statistics, the QML estimation procedure leads to a preference for the AR specifications rather than ARMA however the AR-F model has the lowest AIC value whilst the AR model has the lowest BIC value. In contrast, the MCL estimation procedure prefers the ARMA-F specifications when only considering the AIC values whereas the AR model is preferred under the BIC values. A likelihood ratio test is also employed to determine which model specification is most appropriate. As the AR specification is nested in all other specifications, a standard LR test can be used to test the alternative model over the AR model.

The null hypothesis used to compare the models is $H_0 : \theta = 0, \delta = 0, \kappa = 0$ and choosing an appropriate alternative for each model. Under QML estimation, the LR test statistics for all specifications is insignificant with no p-values above the 10% level. However, under MCL estimation, the ARMA, AR-F and ARMA-F models indicate that the additional terms provide a better model specification as the LR test statistics are significant at the 5% level. Considering the results of the above tests, it can be seen that the choice of estimation procedure can have a marked effect on the choice of model. If QML estimation was used, the AR model would be chosen as there would be no evidence of additional benefit from the other specifications. However, the use of MCL estimation indicates that the inclusion of both an MA term and FOMC meeting dates is appropriate for the modeling of interest rate volatility. However, there is little evidence to suggest that there is any benefit from the inclusion of the term spread as an explanatory variable.

We now turn to the accuracy of the volatility estimates. Following previous studies, the absolute residuals in (10.1) are compared to the conditional standard deviations from the models estimated under both QML and MCL estimation. Given the specifications of the SV models, the conditional volatility is log-normally distributed. However, Harvey and Shephard (1993) argue that as h_t is not known, the expectation of $E(h_{t|T}) - E(h_t)$ is not zero and the assumption of normality for $h_{t|T}$ is not valid. Instead, Harvey and Shephard (1993) suggest that the volatility estimates should be computed as,

$$E(\sigma_t^2) = r_{t-1}^{2\gamma} \exp(h_{t|T}) \qquad (10.11)$$

where $h_{t|T}$ are the smoothed estimates computed using the Kalman filter and smoother.

The MCL conditional mean estimates of the volatility are computed in a similar manner to QML however as MCL uses simulation in the estimation process, the same simulation results can be used to construct the volatility estimates. Under the similar arguments as Harvey and Shephard (1993), Sandmann and Koopman (1998) suggest the following method for computing the posterior mean of the volatility estimates,

$$E(\sigma_t^2) = r_{t-1}^{2\gamma} \exp(h_{t|T}) \qquad (10.12)$$

and,

$$h_{t|T} = \frac{\sum_{i=1}^{N} w(\varepsilon_t^{(i)}) h_t^{(i)}}{\sum_{i=1}^{N} w(\varepsilon_t^{(i)})}, \qquad (10.13)$$

where $w(\varepsilon_t^{(i)})$, $i = 1,...,N$, are the weights used in the estimation of the likelihood function as outlined in the section on MCL algorithm, $h_t^{(i)}$ are the corresponding draws of $g(h|y,\psi)$ and N is the number of simulations used.

Figures 10.2 and 10.3 plot the conditional volatility estimates for both the QML and MCL specifications with the absolute residuals. It is clear from the two sets of figures that there is a big difference between the QML and MCL volatility estimates. Comparing the individual specifications visually, small differences are evident between the AR and ARMA specifications although there is not much distinguishing the specifications.

To provide a better comparison of the different specifications, a number of metrics are used to measure the performance of the models. Similar metrics to previous studies of interest rate volatility are considered including root mean-squared error (RMSE), mean absolute error (MAE), R^2 and Theil's inequality coefficient (TIC). The R^2 is calculated as follows,

$$R^2 = 1 - \frac{\frac{1}{T}\sum_{t=1}^{T}\left(e_t^2 - \hat{\sigma}_t^2 r_{t-1}^{2\gamma}\right)^2}{\frac{1}{T}\sum_{t=1}^{T}\left(e_t^2\right)^2} \tag{10.14}$$

where e_t is the residual from the OLS regression, $\hat{\sigma}_t$ is the estimate of the standard deviation and r_t. Hence the higher the value of the R^2 statistic, the better is the fit. The TIC statistic is computed as follows,

$$TIC = \frac{\sqrt{\frac{1}{T}\sum_{t=1}^{T}\left(e_t^2 - \hat{\sigma}_t r_{t-1}^{\gamma}\right)^2}}{\sqrt{\frac{1}{T}\sum_{t=1}^{T}\left(e_t^2\right)^2} + \sqrt{\frac{1}{T}\sum_{t=1}^{T}\left(\hat{\sigma}_t^2 r_{t-1}^{2\gamma}\right)^2}} \tag{10.15}$$

where the variables are the same as for R^2. This coefficient always lies between zero and one, where the smaller the coefficient the better the estimate. For additional details of this measure and its other applications, readers may refer to Bali and Weinbaum (2006).

Whilst the above measures provide performance measures on the point estimates of the volatility using either the squared residuals or absolute residuals as a benchmark, as neither are a direct measure of volatility, these measures do not give a clear picture of the performance of the models. It may be more informative to determine whether the short rate follows the CKLS process with the volatilities implied by the SV model specifications. One method that can be used to test this is the forecast density test of Berkowitz (2001). The test was inspired by interval forecast tests such as Value at Risk (VaR) but allows for a more in-depth analysis of model performance. The test involves taking the probability integral transform of the observations given the parameter estimates of the model and conducting an LR test on the transformed observations. For the SV model, in section 10.2, e_t was defined as $e_t = \sigma_t r_{t-1}^{\gamma} \varepsilon_t$, which implies that $e_t \mid \sigma_t, F_{t-1} \sim N(0, \sigma_t)$. Hence under the estimates defined by the state space models, the probability integral transform for e_t is given by,

$$p_t = \int_{-\infty}^{e_t} \hat{f}(u, 0, \hat{\sigma}_t^2 r_t^{2\gamma}) du = \hat{F}(e_t) \qquad (10.16)$$

where $e_t \sim N(0, \sigma_{t|T})$ and $\hat{f}(x, \mu, \sigma^2)$ is the normal density function evaluated at e_t with mean μ and variance σ^2. If the smoothed estimates, $\sigma_{t|T}$, implied by the model provide an accurate estimate of volatility, then the series p_t, $t = 1, \ldots, T$, should have a uniform distribution.

However, testing for a uniform distribution is more difficult than testing for normality and Berkowitz (2001) shows that by transforming p_t into standard normal random variables using an inverse normal transform, no loss of information is incurred. It is argued that if the transformation $z_t = \Phi^{-1}(p_t)$ is applied (where $\Phi^{-1}(\cdot)$ is the inverse normal cdf) and the model estimates of the volatility are accurate, it implies that the transformed variables z_t are standard normal. In order to test normality of the transformed series, Berkowitz suggests a standard LR test on z_t by testing the standard normal null against the first-order autoregressive (AR) alternative with mean and variance different from $(0,1)$. For the transformed series, $z_t = \Phi^{-1}(p_t)$, if we define the autoregressive alternative as,

$$z_t - \mu = \rho(z_{t-1} - \mu) + \varepsilon_t \qquad (10.17)$$

then this has a known likelihood function. The likelihood $LR(\mu, \sigma^2, \rho)$ is known to be,

$$
\begin{aligned}
&-\frac{1}{2}\log(2\pi) - \frac{1}{2}\log\left[\sigma^2/(1-\rho^2)\right] - \frac{\left[z_t - \mu/(1-\rho^2)\right]^2}{2\sigma^2(1-\rho^2)} \\
&-\frac{T-1}{2}\log(2\pi) - \frac{T-1}{2}\log(\sigma^2) - \sum_{t=2}^{T}\left[\frac{\left(z_t - \mu - \rho z_{t-1}\right)^2}{2\sigma^2}\right].
\end{aligned}
\tag{10.18}
$$

Hence an LR test statistic can be constructed as,

$$
LR = -2\left[L(0,1,0) - L(\hat{\mu}, \hat{\sigma}^2, \hat{\rho})\right]
\tag{10.19}
$$

which has a $\chi^2(3)$ distribution. Further details of this test can be found in Berkowitz (2001).

The results of the five metrics can be found in Table 10.5. The results for the in-sample tests show that there is little difference between the AR and ARMA specifications of the interest rate volatility, although it indicates that the in-sample fits of the ARMA model perform slightly better. For QML estimation, there is some improvement in the RMSE, MAE and R^2 statistics when comparing the AR specifications with the corresponding ARMA model. The addition of the FOMC meetings as an explanatory variable in the volatility equation appears to increase the accuracy of the estimate whereas the term-spread reduced accuracy.

When considering the results of the forecast density test, it appears that the estimates are not accurate enough under QML as the null is strongly rejected at all traditional levels of significance. Overall, the model performing the best under QML estimation is ARMA-F, although there is not much distinguishing any of the models. The results are somewhat different under MCL estimation. Whilst the metrics show that the addition of the FOMC as an exogenous variable has again improved the accuracy and the term spread has reduced accuracy, the addition of the MA term this time indicates a reduction in accuracy, contrary to the findings in the previous tests.

It can be seen that the metrics evaluating the point estimates support the AR specifications with the AR-F performing the best having the lowest values for the RMSE, MAE, TIC and the highest R^2 value out of all the specifications. However, the results of the point estimate tests may be misleading as the tests use the squared residuals as a proxy for volatility.

As volatility is not directly measurable, testing of point estimates may not give a clear indication on how accurate the estimates are. In this case the forecast density test may be more appropriate as it considers the distribution of the residuals given the volatility estimates rather than evaluating the estimates using a proxy for volatility. Under the forecast density test, it appears that all models fail to reject the null hypothesis at the 5% level indicating that all models under MCL estimation are sufficiently accurate. However, if we consider the values of the LR test statistics, the results appear to indicate a different finding compared to the point estimate tests. In this case, the addition of the MA term improves the estimates slightly, although the addition of both the term-spread and FOMC term shows a slight worsening in performance.

Comparing the above results, it is clear that MCL estimation is superior to QML estimation as expected, supporting the findings of previous authors. However, the evidence also suggests that whilst model selection is important, choice of estimation procedure may be of greater importance. Not only was the improvement in estimation exhibited when using MCL over QML greater than the improvement in the choice of model, the choice of model was also dictated by the estimation procedure.

10.6 Outline of State Space Model Estimation via MCL

Sandmann and Koopman (1998) proposed an extension to the QML method of estimating the SV likelihood function. The method makes use of importance density sampling and is based on Durbin and Koopman (1997) where it is termed Monte Carlo likelihood estimation. Using similar terminology as Durbin and Koopman (1997), for the general state space model,

$$
\begin{aligned}
y_t &= Z_t \alpha_t + d_t + \varepsilon_t \\
\alpha_t &= T_t \alpha_{t-1} + c_t + R_t \eta_t
\end{aligned}
\tag{10.20}
$$

where either ε_t, η_t or both could be non-Gaussian. Letting $Z_t\alpha_t$ be denoted by h_t, the likelihood function is defined as,

$$L(y \mid \psi) = p(y \mid \psi) = \int p(y,h \mid \psi)dh \qquad (10.21)$$

where ψ is a realization of the parameter vector. Using the properties of conditional probability, the likelihood becomes,

$$L(y \mid \psi) = \int p(y \mid h,\psi)p(h \mid \psi)dh . \qquad (10.22)$$

In principle the likelihood can be estimated by using draws of θ but Durbin and Koopman (1997) argue that this is grossly inefficient. Hence they propose by using the fact that a linear Gaussian state space model has the following likelihood function,

$$L_G(y \mid \psi) = g(y \mid \psi) = \frac{g(y,h \mid \psi)}{g(h \mid y,\psi)} = \frac{g(y \mid h,\psi)p(h \mid \psi)}{g(h \mid y,\psi)} \qquad (10.23)$$

and by substituting for $p(\theta)$ in (10.23) into (10.22), the likelihood becomes,

$$\begin{aligned} L(y \mid \psi) &= L_G(y \mid \psi)\int \frac{p(y \mid h,\psi)}{g(y \mid h,\psi)}g(h \mid y,\psi)dh \\ &= L_G(y \mid \psi)E_G\left(\frac{p(y \mid h,\psi)}{g(y \mid h,\psi)}\right) \end{aligned} \qquad (10.24)$$

where E_G is the expectation with respect to the Gaussian importance density and the importance density $g(\theta \mid y,\psi)$ is chosen to be as close to the true density. Hence on comparison of (10.24) and (10.22) it can be seen that the likelihood is computed by its departures from Gaussian rather than estimating the likelihood itself. Also, when the observation disturbances are non-Gaussian and the state disturbances are Gaussian, $p(\varepsilon \mid \psi) = p(y - h \mid \psi) = p(y \mid h,\psi)$ and so the log likelihood is given by,

$$\log L(y \mid \psi) = \log L_G(y \mid \psi) + \log E_G\left(\frac{p(\varepsilon \mid \psi)}{g(\varepsilon \mid \psi)}\right). \qquad (10.25)$$

This shows that the log likelihood can be estimated by computing the Gaussian log likelihood using the standard Kalman filter and a correction for the departures from Gaussian. Sandmann and Koopman (1998) show how to compute the correction term using a simulation smoother. Whilst

Sandmann and Koopman (1998) use the simulation smoother of de Jong and Shephard (1995), the simulation smoother of Durbin and Koopman (2002) is used in this study as this simply allows simultaneous simulation of the state vector if needed. Durbin and Koopman (1997) further show that an unbiased estimate of $\log L(y \mid \psi)$ via simulation smoothing is given by,

$$\log L(y \mid \psi) = \log L_G(y \mid \psi) + \log \overline{\omega} + \frac{s_\omega^2}{2N\overline{\omega}^2} \qquad (10.26)$$

where $\overline{\omega}$ and s_ω^2 are the sample mean and variance of $\omega(\varepsilon^{(i)}) = p(\varepsilon^{(i)} \mid \psi) / g(\varepsilon^{(i)} \mid \psi)$ and $\varepsilon^{(i)}$ is a draw from the Gaussian importance density $g(\varepsilon^{(i)} \mid y)$. The estimates of the parameter vector are obtained through numerical maximization of the log likelihood function in (10.26).

10.7 Concluding Summary

Our investigation of interest rate stochastic volatility considered a number of different specifications for the log-variance equation in the class of discrete-time SV models using both QML and MCL estimation procedures. Some similar results were exhibited by the models where it was found that whilst persistence in volatility was exhibited by the short rate process, it was not as high as is usually found in stocks. Also, a crucial parameter in interest rate volatility models is the elasticity parameter. The elasticity gives a clear indication of how interest rate volatility varies with the level of interest rates. The results found in this paper support the CIR model of interest rates indicating that an elasticity parameter of 0.5 is appropriate and the value of around 1.5 found in CKLS is too high as it does not account for time-varying volatility.

Our examination also included introducing an MA term, a yield term spread and FOMC meeting dates in the log-variance equation. Using standard model selection criteria, the empirical results indicate that the inclusion of both an MA term and the dates of FOMC meetings should be considered when modeling interest rate volatility, although the results were inconclusive as to whether the term spread should be an included variable. The evidence suggests that FOMC meetings increase volatility suggesting that the meetings provide information to market participants resulting in an increase in trading activity and hence volatility.

However, it was found that the choice of estimation procedure can be of more importance than model selection when estimating SV models. The choice of an efficient estimation procedure should be considered before considering alternative SV models as the estimation procedure may also dictate the choice of model.

The author has also extended this analysis to using Particle Filter and then compared all these three methods i.e. QML, MCL and Particle Filter in a working paper available at: http://ssrn.com/abstract=1160659.

Table 10.1 Summary Statistics Three-month Constant Maturity U.S. Treasury Yields and Residuals of the AR(1) Model.

	r_t	$e_t = \Delta r_t - a - b r_{t-1}$
Mean	0.0419	0.0000
Standard Deviation	0.018	0.001
Skewness	-0.156	-0.680
Kurtosis	2.398	10.889
Jarque-Bera	16.39	2283
	(0.000)	(0.000)
Portmanteau	26270.631	166.156
	(0.000)	(0.000)
Augmented Dickey-Fuller	-1.640	-31.332
	(0.095)	(0.000)
ARCH LM	844.673	127.120
	(0.000)	(0.000)

p-values for all test statistics are in parentheses. Statistical significance is tested at the 5% level.

Table 10.2 Parameter Estimates of the Mean Equation.

a	0.0001
	(0.000)
b	-0.0040
	(0.002)

Parameters are estimated by OLS where $\Delta r_t = a + b r_{t-1} + e_t$.

Table 10.3 QML Estimated Model Parameters: SV Model for Short-Term Interest Rate (USA).

	AR	ARMA	AR-S	ARMA-S	AR-F	ARMA-F
μ	-6.913	-6.920	-7.019	-7.066	-6.978	-6.984
	(0.091)	(0.093)	(0.306)	(0.312)	(0.092)	(0.101)
φ	0.608	0.640	0.604	0.639	0.598	0.632
	(0.016)	(0.015)	(0.017)	(0.015)	(0.017)	(0.016)
θ	-	0.084	-	0.117	-	0.096
	-	(0.045)	-	(0.047)	-	(0.046)
σ_η	0.863	0.739	0.868	0.715	0.875	0.737
	(0.046)	(0.002)	(0.047)	(0.002)	(0.049)	(0.002)
γ	0.657	0.657	0.677	0.685	0.656	0.656
	(0.011)	(0.011)	(0.020)	(0.020)	(0.011)	(0.011)
δ	-	-	0.031	0.042	-	-
	-	-	(0.020)	(0.018)	-	-
κ	-	-	-	-	0.421	0.422
	-	-	-	-	(0.056)	(0.055)
Loglik	-1994.6	-1994.4	-1994.6	-1994.3	-1992.9	-1992.7
AIC	3999.3	4000.8	4001.2	4002.7	3997.8	3999.3
BIC	4023	4029.3	4029.7	4035.9	4026.3	4032.6
LR Test	-	0.6	0	0.8	3.4	4
	-	(0.4386)	(1.0000)	(0.3711)	(0.0652)	(0.1353)

Model parameters are estimated by the quasi-maximum likelihood method using a numerical maximization algorithm in Matlab. The robust standard errors are shown in parentheses.

Table 10.4 MCL Estimated Model Parameters: SV Model for Short-Term Interest Rate (USA).

	AR	ARMA	AR-S	ARMA-S	AR-F	ARMA-F
μ	-6.742	-6.754	-7.043	-7.130	-6.778	-6.763
	(0.053)	(0.051)	(0.171)	(0.17)	(0.052)	(0.060)
φ	0.674	0.597	0.668	0.594	0.676	0.641
	(0.003)	(0.007)	(0.003)	(0.006)	(0.004)	(0.014)
θ	-	0.620	-	0.629	-	0.530
	-	(0.337)	-	(0.338)	-	(0.445)
σ_η	0.694	0.517	0.696	0.512	0.675	0.493
	(0.004)	(0.028)	(0.003)	(0.028)	(0.004)	(0.065)
γ	0.598	0.595	0.657	0.676	0.595	0.595
	(0.007)	(0.006)	(0.011)	(0.011)	(0.006)	(0.007)
δ	-	-	0.087	0.109	-	-
	-	-	(0.010)	(0.010)	-	-
κ	-	-	-	-	0.357	0.348
	-	-	-	-	(0.024)	(0.027)
Loglik	-1900.3	-1898.2	-1899.9	-1897.6	-1897.4	-1895.8
AIC	3808.5	3806.4	3811.8	3809.3	3804.8	3803.6
BIC	3827.5	3830.2	3840.3	3842.5	3828.6	3832.1
LR Test	-	4.2	0.8	5.4	5.8	9
	-	(0.0404)	(0.3711)	(0.0672)	(0.0160)	(0.0111)

Model parameters are estimated by the Monte Carlo likelihood method using a numerical maximization algorithm in Matlab. The robust standard errors are shown in parentheses.

Table 10.5 Relative Performance of Models.

	AR	ARMA	AR-S	ARMA-S	AR-F	ARMA-F
QML						
RMSE	0.0368	0.0368	0.0369	0.0369	0.0367	0.0367
MAE	0.0101	0.0101	0.0101	0.0101	0.0101	0.101
R^2	0.8835	0.8835	0.8834	0.8834	0.8846	0.8844
TIC	0.0329	0.0329	0.0330	0.0330	0.0325	0.0326
FD-LR	1255.34	1225.59	1248.13	1218.42	1163.84	1137.93
	(0.0000)	(0.0000)	(0.0000)	(0.0000)	(0.0000)	(0.0000)
MCL						
RMSE	0.0264	0.0267	0.0265	0.0269	0.0255	0.0265
MAE	0.0101	0.0102	0.0101	0.0103	0.0102	0.0104
R^2	0.9404	0.9388	0.9399	0.9379	0.9440	0.9400
TIC	0.0121	0.0125	0.0123	0.0127	0.0112	0.0122
FD-LR	5.5605	4.9872	6.5351	5.4383	5.6769	5.0361
	(0.1351)	(0.1727)	(0.0883)	(0.1424)	(0.1284)	(0.1692)

RMSE is defined as $\sum_{t=1}^{T}(e_t^2 - \sigma_t^2)/T$.

MAE is defined as $\sum_{t=1}^{T}\left|e_t^2 - \sigma_t^2\right|/T$.

R^2 is described in the text where a value close to 1 implies a good fit. TIC is Theil's inequality coefficient and it is described in the text. This coefficient lies between 0 and 1. For a perfect forecast its value is 0, hence a smaller value implies a better model. FD-LR is the forecast density likelihood ratio test statistic. The statistic is a $\chi^2(3)$ distributed random variable with p-values in parentheses.

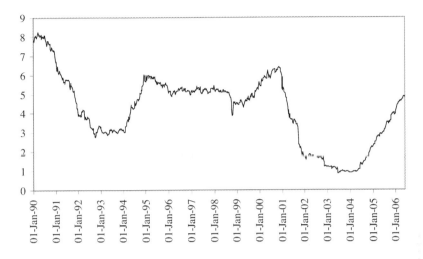

Figure 10.1(a) Yields on 3-month constant maturity bonds.

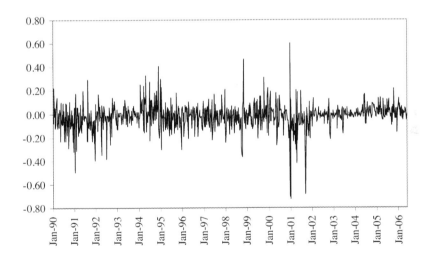

Figure 10.1(b) Residuals of the CKLS discretization.

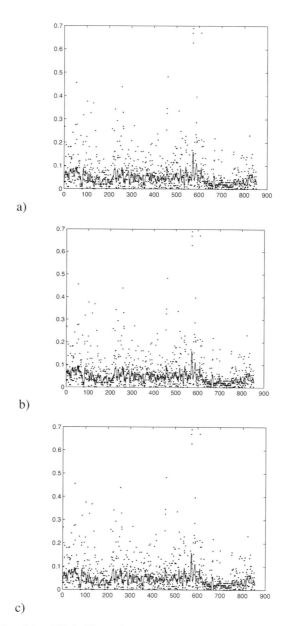

Figure 10.2 Conditional Volatility estimates under QML estimation: (a) AR; (b) ARMA; (c) AR-S.

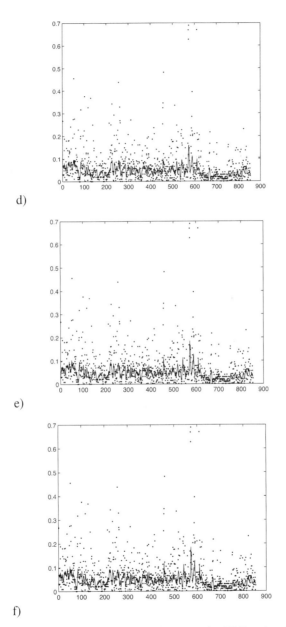

Figure 10.2 (Contd.) Conditional Volatility estimates under QML estimation: (d) ARMA-S; (e) AR-F; (f) ARMA-F.

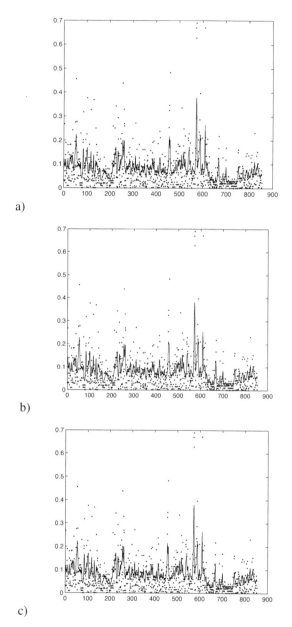

a)

b)

c)

Figure 10.3 Conditional Volatility Estimates under MCL estimation: (a) AR; (b) ARMA; (c) AR-S.

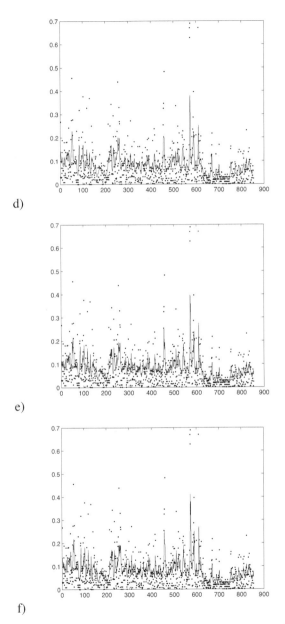

Figure 10.3 (Contd.) Conditional Volatility Estimates under MCL estimation: (d) ARMA-S; (e) AR-F; (f) ARMA-F.

Chapter 11

Applications for Filtering with Jumps

So far the models considered are mainly described via some form diffusion process. There might have been non-linear relationship between the unobserved driving components and the measurable observations. In this chapter we focus on those assets and/or characteristics that are influenced by occasional jumps. In other words, these processes are described by both jumps and diffusions. The commodity market and in particular the electricity market is a prime example where occasional large jumps make the models difficult to tract. Firstly, in this chapter we discuss a suitable model for the spot electricity prices in the state of New South Wales in Australia and show how one step ahead price forecast may be made. Secondly, we show that CDS (Credit Default Swap) prices from the European market also exhibit jumps. This raises the question whether such jump risks could be diversified by some way. Both these cases require stochastic filtering that incorporates jumps and such jumps are defined by suitable Poisson variables.

In the first application of evolution of spot electricity prices in New South Wales, the model allows a deterministic time trend component and an unobserved process driven by both time varying volatility and occasional jumps. The structure allows us to cast the problem in a state space form and suitable modification of the Kalman filter enables us to infer the unobserved driving process. The one-step ahead predicted price based on this component structure performs reasonably well in capturing the patterns in the daily average spot prices. This work may be viewed as an initial attempt and much more work in the area needs to be undertaken.

In the second application the focus is on Credit Default Swaps (CDS) index as well sub indices with a view to diversify jump risks. Previous research on credit risk mainly focuses on the prediction of default probability and most of these are based upon bond market analyses. The rapid-development of the CDS market makes research on credit risk using information from this market more important and attractive. Also, over the last few years, analysis of dynamic behaviour of credit risk has become important among academics, practitioners, and regulators. The key question being investigated here is non diversifiable jump risk in iTraxx sector indices based on a multivariate framework that explicitly admits discrete common jumps for an index and its components. Our empirical research shows that both the sector iTraxx indices and the iTraxx Non-Financials, which acts as an index for the sector, experience jumps during sample period. This means that the jump risks in iTraxx sector index are not diversifiable. Such a result warrants further investigation on credit risk, for example, in pricing and prediction.

11.1 Background to Electricity Market and Prices

At the outset, to appreciate the background of this project, we quote Geman and Roncoroni (2006):

"Over the last 10 years, major countries have been experiencing deregulation in generation and supply activities. One of the important consequences of this restructuring is that prices now determined according to the fundamental rule of supply and demand: there is a 'market pool' in which bids are placed by generators to sell electricity for the next day are compared to purchase orders."

Prior to that, the regulators used to set the price based on the cost of generation, transmission and distribution and the price to the consumer was essentially fixed for long period of time. A large fraction of the literature on electricity today belongs to the economics of deregulated electricity market from the perspective of the regulators (see Joskow and Kahn (2001)). In the market mechanism now operating today, the price will be determined by the interaction of the purchase orders placed by the retailers against the pool prices.

The deregulation of the electricity market has also led to increased trading activities in both spot and related derivatives like forwards and options. The risk of spot-price has forced retailers to manage the risk of the spot price through various hedging mechanisms. Many retailers provide incentives to the consumers to enter into long term contract with pre-determined price structure, but that still leaves the risk of buying price. It is in this context modeling the stochastic behavior of the spot price of electricity has become important.

One feature of the electricity market that is unique to this commodity is that electricity is not storable, although, it may be argued that the concept of storability applies to hydro electricity generation. Since, in general, it cannot be stored the spot price is likely to be determined by the spot concerns, e.g., spot demand and supply constraints. The ability to store any commodity has the effect of smoothing the evolution of the spot price to some extent. As a result of its absence, price spikes are a regular feature of the electricity spot prices in most countries that have deregulated this market. Price spikes are possibly due to disruption in transmission, unscheduled outages, extreme weather changes or a combination of all these events. Additional details about the characteristics of this market may be found in Geman and Roncoroni (2006).

We will now review some of the salient characteristics of the electricity prices in the deregulated market. In standard commodity-futures markets the concept of convenience yield plays a key role in the relationship between the spot and the forward prices. The convenience yield is a way of expressing the fact that an investor is sure of available supply when the demand for using that commodity arises at a future date. The non-storability of electricity makes the concept of convenience yield difficult to apply. This implies that the spot price itself should contain all the characteristics of the price process that would be necessary to impute prices of derivatives contracts written on electricity prices.

Next, we outline the important temporal characteristics of spot electricity prices observed in most markets. A detailed description of these characteristics may be found in Geman and Roncoroni (2006).

Mean reversion is an important feature of spot electricity prices. The prices tend to fluctuate around values determined by cost of production and the level of demand. The mean reversion level may be constant or periodic with a trend. Seasonality is another obvious characteristic. The prices change by time of day, week, month and the year in response to cyclical fluctuations in demand. Another feature already mentioned before is that price jumps or spikes. A point to note is that technically price does not jump to a new level (to stay there) but spikes and quickly reverts to their previous levels. This price spike has been the most difficult aspect from modeling purposes.

It is, therefore, clear from the above discussions that a pure diffusion process would not adequately capture the characteristics for electricity price series. A pure diffusion process approach has worked well in stock price modeling. For the electricity market, however, we need to incorporate a jump component with an appropriate intensity function to capture the spikes. Many of the traditional modeling approaches applied to financial market data e.g. equity, foreign exchange, and interest rates etc do not work well with spot electricity prices. This has been the experience for most researchers in this area as discussed in Geman and Roncoroni (2006). With respect to the equity market though, the work by Kim, Oh and Brooks (1994) is an important contribution to detect jumps (as opposed to spike). Their focus has been whether jump risks in stock returns are diversifiable.

In this chapter we attempt to combine the ideas expressed in the cited literatures and explore a jump diffusion model for spot electricity prices in NSW. We allow both a deterministic time dependent factor as well as a latent factor combined with Poisson jumps to capture the observed characteristics of the spot electricity price series. We show how to calibrate such a model to the market data and describe the appropriate algorithm for that. The algorithm we employ generates, in a natural way, one period ahead forecast of spot electricity price. This, in turn, helps us determine the "goodness of fit" of such a model.

11.2 A Model for Spot Electricity Prices

In modeling commodity prices the approach of Schwartz and Smith (2000) has become quite popular. Their analyses depend upon both short dated and long dated futures contracts of the commodity. It also relate to the convenience yield as normally applied to futures contracts. Since the electricity, as a commodity, is different in this respect due to non-storability of the commodity for possible future consumption, the short-term, long-term concept introduced by Schwartz and Smith may not strictly apply to this market. Nevertheless, the ideas contained in Schwartz and Smith has important bearing in dealing with the electricity market.

It is clear from the earlier discussions that price jumps or price spikes are a natural characteristic of the electricity market and have to be built into the model. It is also useful as we may be able to adopt the models we develop here for pricing derivatives contracts on the electricity spot prices. To reliably model contingent claims prices we have to incorporate jumps in addition to the usual diffusion assumptions in the price process, which makes it far more complex compared to pricing derivatives on equities. In this context we need to be mindful of the theorem by Duffie, Pan and Singleton (2000) that leads to closed form solution, in most cases, of the contingent claims when the underlying security follows an affine jump-diffusion process (AJD) (affine structure implies linear dependence on state variables). Although, we are not strictly focussing on electricity derivatives contract in this paper, we will strive to stay close to the AJD process so that our approach can be easily adapted for contingent claims pricing later.

Many researchers traditionally model log of the spot price of the commodity as in Schwartz and Smith (2000). The existence of a significant jump component in the electricity prices it is worthwhile to re-consider whether a logarithmic transformation is useful. The logarithmic transformation affects the estimation of the jump component due its effect on the skewness of the distribution of the series. Since the derivatives contracts are written on spot price level and not on its log transformation, developing models of log transformation of spot price will not be useful. Lucia and Schwartz (2002) find that the model of price level fits the forward contract prices better than the log-price level. In this paper we will, therefore, model the price level and not its log transformation. That way the models we develop will be better suited to pricing derivatives contracts on spot electricity prices.

In the original approach of Schwartz and Smith (2000) the log of the commodity price is modeled via two factors, both unobserved. The first factor captures the short-term variations and is modeled by an Ornstein-Uhlenbeck (OU) process whereas the second factor (the long-term variations) is modeled by an Arithmetic Brownian process (ABM). The commodity examined in Schwartz and Smith is crude oil and it display non-stationarity. Hence the inclusion of the ABM process in their analysis is not only meaningful but is also a necessity since the OU process alone would not be able to capture the dynamics. Since our spot electricity price series is stationary (found by Augmented Dickey-Fuller tests) we need only include the OU process to capture the dynamics without the jumps. To capture the jump characteristic we include a jump component in the OU process.

Another difference from the structure in Schwartz and Smith (2000) for electricity spot prices is the inclusion of a time-dependent, deterministic function to capture the observed seasonality in the series. This arises mainly due to the nature of household consumptions of electricity depending on the season we are in. This also indicates that the intensity process for the Poisson component capturing the jumps in the series may not be constant, and is more likely to depend on seasonal factors.

With this background we are in a position to specify the spot price process (P_t) mathematically in terms of a deterministic, time-dependent function $f(t)$, and a state variable X_t. Although many researchers specify their models in continuous time setting and for implementation purposes use Euler discretisation, we prefer to stay in the discrete framework from the start. We set daily average electricity spot price, measured in dollars per megawatt-hour,

$$P_t = f(t) + X_t. \tag{11.1}$$

The deterministic, time-dependent part is described as a sinusoidal function along with a weekday dummy variable (wkd_t). This specification is similar to that used in Lucia and Schwartz (2002). This last variable is to help distinguish between the price on a weekday and a weekend. We use the following specification for $f(t)$:

$$f(t) = \beta_0 + \beta_1 t + \beta_2 \sin\left\{(t + \beta_3)\frac{2\pi}{365}\right\} + \beta_4 \sin\left\{(t + \beta_5)\frac{4\pi}{365}\right\}$$
$$+ \beta_6 wkd_t \qquad (11.2)$$

where $wkd_t = 1$ if the day is a weekday otherwise it is zero.

The OU component X_t describes the unobserved component that captures the short-term dynamics, allows occasional jumps and is characterized by volatility clustering that is common to most financial time series. The notion of unobserved component is in the sense of state space models and filtering theory. This volatility clustering is modeled as a GARCH $(1,1)$ process. Thus we specify:

$$X_t = \phi X_{t-1} + h_t^{0.5}\varepsilon_{1,t} + J(\mu_J, \sigma_J)\Delta\Pi(\lambda_t) \qquad (11.3)$$

where $\varepsilon_{1,t} \sim N(0,1)$, the time varying variance $h_t = \alpha_0 + \alpha_1 \varepsilon_{1,t-1}^2 + \alpha_2 h_{t-1}$. The jump component is controlled by a Poisson-distributed variable with time dependent intensity function λ_t, and the jump amplitude is governed by a normally distributed variable with mean μ_J and variance σ_J. To capture seasonal effects in the jump component, we set

$$\lambda_t = \gamma_1 winter_t + \gamma_2 autumn_t + \gamma_3 spring_t + \gamma_4 summer_t, \qquad (11.4)$$

where the seasonal dummy variables indicate whether a particular date is in one of these seasons.

The specifications in equations (11.2) and (11.4) are just an assumption of the form of these functions for the dataset being analyzed. These are based upon basic analysis of the data and similar decisions taken in the literature already referred to above. In this analysis we have not explored whether other forms of these functions have better properties.

The set of equations (11.1) and (11.3) describe our modeling approach to electricity spot prices and is already in state space form. In this state space representation of our problem equation (11.1) is the measurement equation and equation (11.3) is the state transition equation. The number of unknown parameters in this model is 17 and these are estimated by the maximum likelihood method. Due to the presence of the unobserved component, X_t, we resort to Kalman filter to (as described in Chapter 1) develop the likelihood function recursively. This process along with the modification needed due to the jump process, is described in the next section. The parameter set is conveniently given by the vector,

$$\Theta \equiv \{\beta_0, \beta_1, \beta_2, \beta_3, \beta_4, \beta_5, \beta_6, \alpha_0, \alpha_1, \alpha_2, \mu_J, \sigma_J, \gamma_1, \gamma_2, \gamma_3, \gamma_4, \phi\}. \quad (11.5)$$

There are seventeen parameters to be estimated and it is not an easy optimization task.

Although, we have described in the next section in detail the steps of the filtering algorithm including the modification needed to accommodate Poisson jumps, additional insights may be gained from chapter 6 in Kim and Nelson (1999). In order to allow GARCH variance in the state dynamics, we need further non-trivial modification to the standard Kalman filter. Kim and Nelson (1999), and in particular Chapter 6 is an excellent reference source for this topic. Thus to conserve space, we refer the reader to that source material.

This algorithm may be implemented in Gauss or Matlab programming environment and numerical optimization is needed to estimate the parameters as well as the standard errors from the information matrix. The filter algorithm produces the one-step ahead prediction of the state vector, which in this case is X_t.

11.3 State Space Model, Kalman Filter and Poisson Jumps

The state space model (SSM) in its basic form retains a VAR (1) structure for the state equation,

$$y_t = \Gamma y_{t-1} + w_t + \varsigma_t \quad (11.6)$$

where the state equation determines the rule for generation of the states y_t, $p \times 1$ vector, from the past states y_{t-1}, for all time points $t = 1, 2, ..., T$. For completeness we assume that w_t are $p \times 1$ independent and identically distributed zero-mean normal vectors with covariance Q_t. We assume that Q_t is diagonal and may be constant.

When we allow GARCH effect in some of the elements of the state vector, the corresponding element in Q_t would then be time varying. The noise term ς_t introduces the jump in the process and is assumed to be represented by,

$$\varsigma_t \sim N\left(j \cdot \mu_j, j \cdot \sigma_j^2\right),\ j = 1, 2, \ldots, \infty.\qquad(11.7)$$

In equation (11.7) j is a Poisson distributed random variable during a small interval Δt characterized by a single parameter $\lambda_t \Delta t$. The state process is assumed to have started with the initial value given by the vector, y_0, taken from normally distributed variables with mean vector μ_0 and the $p \times p$ covariance matrix, Σ_0.

The state vector itself is not observed but some transformation of these is observed but in a linearly added noisy environment. Thus, the measurement equation is given by,

$$z_t = d_t + A_t y_t + v_t.\qquad(11.8)$$

In this sense, the $q \times 1$ vector z_t is observed through the $q \times p$ measurement matrix A_t together with the $q \times 1$ Gaussian white noise v_t, with the covariance matrix R. In equation (11.8) d_t is a purely deterministic time dependent variable. We also assume that the two noise sources in the state and the measurement equations are uncorrelated.

The next step is to make use of the Gaussian assumptions and the independence of Poisson distributed events and across times and produce estimates of the underlying unobserved state vector given the measurements up to a particular point in time. In other words, we would like to find out, $E\left(y_t \mid \{z_{t-1}, z_{t-2} \cdots z_1\}\right)$ and the covariance matrix, $P_{t|t-1} = E\left[\left(y_t - y_{t|t-1}\right)\left(y_t - y_{t|t-1}\right)'\right]$. This is achieved by using Kalman filter and the basic system of equations is described below for this section to be self contained for the readers.

Given the initial conditions $y_{0|0} = \mu_0$, and $P_{0|0} = \Sigma_0$, for observations made at time $1, 2, 3 \ldots T$,

$$y_{t|t-1}^{(j)} = \Gamma y_{t-1|t-1}^{(j)}\qquad(11.9)$$

$$P_{t|t-1}^{(j)} = \Gamma P_{t-1|t-1}^{(j)} \Gamma' + Q_t\qquad(11.10)$$

$$y_{t|t}^{(j)} = y_{t|t-1}^{(j)} + K_t^{(j)}\left(z_t - A_t y_{t|t-1}^{(j)}\right)\qquad(11.11)$$

where the Kalman gain matrix,

$$K_t^{(j)} = P_{t|t-1}^{(j)} A_t' \left[A_t P_{t|t-1} A_t' + R \right]^{-1} \tag{11.12}$$

and the covariance matrix $P_{t|t}$ after the tth measurement has been made is,

$$P_{t|t}^{(j)} = \left[I - K_t^{(j)} A_t \right] P_{t|t-1}^{(j)} . \tag{11.13}$$

Equation (11.9) forecasts the state vector for the next period given the current state vector and the Poisson jump. Using this one step ahead forecast of the state vector it is possible to define the innovation vector as,

$$v_t^{(j)} = z_t - A_t y_{t|t-1}^{(j)} \tag{11.14}$$

and its covariance as,

$$\Sigma_t^{(j)} = A_t P_{t|t-1}^{(j)} A_t' + R . \tag{11.15}$$

The description of the above filtering algorithms assumes that the parameters are known. In fact, we want to determine these parameters and this is achieved by maximizing the innovation form of the likelihood function. The one step ahead innovation and its covariance matrix are defined by the equations (11.14 and 11.15) and since these are assumed to be independent and conditionally Gaussian, the log likelihood of the Poisson mixture of normal distribution is given by,

$$Ln(L) = \sum_{t=1}^{T} Ln \left[\sum_{j=0}^{\infty} \omega(\lambda_t, j) \left\{ (2\pi)^{-\frac{q}{2}} \cdot \left| \Sigma_t(\Theta) \right|^{-\frac{1}{2}} \times \Psi \right\} \right] \tag{11.16}$$

$$\Psi \equiv \exp\left(-\frac{1}{2} v_t^{(j)'}(\Theta) \Sigma_t^{(j)-1}(\Theta) v_t^{(j)}(\Theta) \right).$$

In this expression Θ (collection of all the unknown parameters) is specifically used to emphasize the dependence of the log likelihood function on the parameters of the model. Once the function is maximized with respect to the parameters of the model, the inferred state vector is also available.

In practice the infinite sum in the above log likelihood function has to be approximated by something more appropriate for computation. The published papers in this area using equity market data normally use 10 as an upper limit for the summation term. For example, Kim, Oh and Brooks (1994) use 4 for their study of jump risks in equity return. In this paper we use 6. Higher value will of course give better approximation but at the expense of rapidly increased computation time.

There are different numerical approaches that may be taken to carry out the maximization of the log likelihood function. The computational complexity and other numerical issues are beyond the scope of this paper.

11.4 Data and Empirical Results for Electricity Market

Data used for this study was provided by a local retail electricity supplier. This represents every half hour NSW pool price covering the period 2002 through to 2006. We, however, use the daily average price for our modeling. We also estimate the model for two different periods. The first sample covers 2002–2003 containing 730 observations and the second sample covers 2004–2006 containing 1096 observations. Daily average prices are expressed in Australian dollars.

As a first step we check the stationarity of the spot price series using an ADF test. The main focus here is to statistically reject the unit root hypothesis in the electricity spot price series for entire sample period examined. The ADF test is based upon the following equation, where P_t is the daily average electricity spot price and Δ is the difference operator:

$$\Delta P_t = c + \gamma P_{t-1} + \sum_{i=1}^{n} \delta_i \Delta P_{t-i} + \eta_t \,. \qquad (11.17)$$

The hypothesis being tested is $H_0 : \gamma = 0$ as opposed to $H_1 : \gamma < 0$. The quantity n in the above test is decided by sample specific check so that the residual series is uncorrelated. The t-statistic for this test for the sample of 2002–2003 is -13.52 and that for the sample 2004–2006 is -27.04. The critical values for this test are obtained from the econometric software Eviews^{TM} and the existence of unit root is rejected for both samples.

This unit root test convincingly supports the view that we need only the OU component in our model as opposed to Schwartz and Smith

(2000) where both an OU and an ABM component were needed. For our analysis this OU component is given by the equation (11.3).

Focusing on the parameter estimates in Table 11.1, we note that the time trend component in the deterministic part (equation (11.2)) is insignificant in the second sample (2004–2006), whereas in the first component it is highly significant and displays downward bias. Although we cannot draw any firm economic conclusion from this, it is worthwhile to keep in mind that this market is still evolving and maturing.

Both samples display heteroscedasticity as seen from the quantity, $(\alpha_1 + \alpha_2)$. The persistence in volatility in the short-term component (the OU part) has, however, gone down in the second sample. It is also interesting to observe that the autoregressive parameter (ϕ) of the unobserved component in the second sample is about half the size of the earlier sample. It may stem from more efficient pricing by the participants in this market and is probably the result of better understanding of this commodity in this evolving market.

We now focus on the jump component, i.e., the parameters (μ_J, σ_J). Both these parameters are highly significant in both samples. Although the mean jump amplitude is higher in the second sample, its volatility is an order of magnitude higher in the second sample. That it might be so is also apparent from the Figures 11.1 and 11.3 which are plots of the electricity spot prices over the whole periods. This may result from the supply concerns in the second sample period or the reflection of uncertainty in the regulatory environment governing this market. In this analysis we are not in a position to shed further light on this aspect of the results. The appropriateness of the time varying jump intensity as captured by equation (11.4) is supported by the statistical significances of the estimated parameters in both sample periods.

Finally, we check on the predictive power of the model in both sample periods. Since the filtering algorithm recursively produces one step ahead projection of the state variable, we have shown in Figures 2 and 4 the possible price paths i.e. the expected electricity spot prices. Using this information and the subsequently realized prices we can make comparative judgment about the usefulness of the model. In the traditional statistical sense the forecast ability of a model is judged by some measure of association between the forecasts and the realizations.

However, there is an alternative to R-square measures and this is given by Theil's Inequality Coefficient. It was originally proposed in 1961 and has been employed by several researchers since then. We have applied Theil's Inequality Coefficient (TIC) to test this performance.

The TIC (Theil's Inequality Coefficient is given by the following expression. We assume that the variable of interest is z_t for t=1,2,...T, and its estimated value is given by $\hat{z}_t$:

$$\text{TIC} = \frac{\sqrt{1/T\sum_{t=1}^{T}(z_t - \hat{z}_t)^2}}{\sqrt{1/T\sum_{t=1}^{T} z_t^2} + \sqrt{1/T\sum_{t=1}^{T} \hat{z}_t^2}} . \qquad (11.18)$$

This coefficient always lies between zero and one, where the smaller the coefficient the better the estimate. For additional application of this measure the readers may refer to Bali and Weinbaum (2007).

The computed value of TIC is smaller in the first sample compared to that in the second sample. Both values, however, indicate reasonable success in capturing the price path one day ahead. The lower value of the autoregressive parameter (ϕ) in the second sample may have contributed to lower predictive power in the second sample.

11.5 Summarizing Electricity Market Application

We have explored the modeling of the electricity spot prices in New South Wales through a jump-diffusion process mixed with time varying deterministic trend component. The unobserved factor is driven by a diffusion process with time varying variance and a Poisson distributed jump component. We have outlined the algorithm needed to extract this latent factor from the observed price series and find that its one-step ahead prediction does contribute to the forecast of a complex electricity spot price series.

The very nature of the likelihood function of the Poisson mixture of Gaussian distribution requires some approximation for implementation purposes. To keep the computation burden low we have kept the upper limit of the infinite series summation to a small value. This is consistent with published articles in this area of research. One way to extend this study would be to experiment with this upper limit to check whether it improves predictive accuracy. Although we have allowed the jump amplitude to be a normally distributed variable, there may be other distributions that could prove useful in improving the predictive accuracy. This remains another possible extension of this study.

11.6 Background to Jumps in CDS Indices

The topic of Credit Default Swaps (CDS) has been covered in detail in Chapters 8 and 9. This chapter focuses on another aspect of the CDS dealing with jump risks in the index and in the components as well as the ability or otherwise to diversify this risk. Before proceeding, however, the salient aspects are reviewed for this important development in the credit market.

CDS is commonly cited as a basic "building block" for more complex structures. A single-name CDS is a contract that protects the buyer, who is the holder of an underlying obligation, from the losses caused by the occurrence of a credit event to the obligation's issuer, referred to as the reference entity. The protection buyer pays periodically to the protection seller, typically a percentage of the notional amount. The percentage that gives the contract zero value at initiation is called the spread. When credit event occurs, the protection buyer is also required to pay the part of the premium payment that has accrued since the last payment date, called accrual payment.

The reasons that the CDS's are more attractive than credit spreads, in credit risk research, are summarized as follows. Firstly, CDS spreads do not require the specification of a benchmark risk free yield curve as does the credit spread. The credit spread comprises both a premium for bearing default risk and a premium for bearing interest rate risk. As Longstaff, Neis and Mithal (2005) point out, different benchmark risk-free rates imply different magnitudes for the default component in credit spreads. Secondly, short-sale constraints in corporate bond markets mean that credit spreads are not a 'pure' measure of credit risk either. Joost (2005) finds that the liquidity premium accounts for 17% to 25% of the bond yield over Treasuries. Furthermore, Ericsson and Renault (2006) demonstrate that as default becomes more likely, the components of credit spreads (also called bond yield spreads) attributable to illiquidity increase. Thirdly, counterparty risk generally does not need to be taken into account in CDS spreads. This is because, if a party is unwilling to take on credit risk to its counterparty, it decides either to cancel the trade or to alleviate the exposure. For example, it is possible to demand that collateral is provided or that the premium is paid up-front instead of periodically, Culp and Neves (1998).

With the development of the CDS market, similar to the way a stock index is created as a portfolio of individual stocks, a CDS index, for example iTraxx is a equally weighted portfolio of single-name credit default swaps. The iTraxx index family consists of various indices of the most liquid CDS contracts in Europe and Asia (in the US, since April 2004, a similar family of indices is called Dow Jones CDX). The iTraxx indices typically trade 5 and 10-year maturities and a new series is determined by dealer liquidity poll every 6 months. The Europe and HiVol indices also trade 3 and 7-year maturities. The indices are managed and administrated by a newly created company called International Index Company that is owned by a group of the largest global investment banks.

The benchmark iTraxx Europe index comprises 125 equally-weighted European names which are selected by a dealer poll based on CDS volume traded over the previous six months. The European index is further split up into several sector indices (Non-financials, financials et.), a corporate index comprised of the largest non-financial names (from the 125 names), a crossover index comprising the 25 most liquid sub-investment grade non-financial names and a HiVol index that consists of the 30 names with the widest CDS spreads. Table 11.2 gives an overview of these iTraxx index families.

Given the attraction of CDS spread data and the importance of dynamic credit risk modeling, we investigate in this chapter a multivariate model for its dynamic behavior and possible unobservable common factor behind iTraxx sector indices data. That is, we chose iTraxx Non-Financials and its sub-indices, namely, iTraxx TMT, iTraxx Industrial, iTraxx Energy, iTraxx Autos, and iTraxx consumers. All index quotes have been made available by the International Index

Company. In this analysis we only present results for the 5-year maturity iTraxx indices. However, similar results can be obtained form those of the 10-year maturities.

There are two basic approaches to modeling credit risks, structural models and reduced form model. Structural models, pioneered by Black and Scholes (1973) and Merton (1974), have been extended by Black and Cox (1976), and Longstaff and Schwartz (1995a) among others. In this approach, default occurrence is modeled when the value of the firm hits a given boundary. Reduced form models, on the other hand, developed by Litterman and Iben (1991), Jarrow and Turnbull (1995) and Jarrow et al. (1997), assume that credit risk is determined by the occurrence of default and its recovered amount. Default is often represented by a random stopping time with stochastic or deterministic arrival intensity (hazard rate). Most often in these models the recovery rate is assumed to be constant.

Classical structural models are typically based on diffusion processes where default occurs only when the firm's value reaches some exogenous pre-specified barrier that can be time-dependent. On the other hand, credit markets, even in the short term, do not rule out the possibility that a firm can default instantaneously because of a sudden drop in its value. Such a phenomenon has been identified in empirical work as well. For example, Jones, Mason, and Rosenfeld (1984) find that the credit spreads on corporate bonds are too high to be matched by the model based on the diffusion process.

The particular focus in this analysis is on the Jump risk in our sample of iTraxx indices. Jump diffusion processes have been applied to stock prices and commodity prices. For example, Myung-Jig et al (1994) record non-diversifiable jump risk in stock returns. They present a multivariate time-series model of the stochastic process that explicitly admits discrete common jumps for an index and its component stocks. However, not much work has been done directly investigating the CDS spreads to examine jump risks. They found that the Poisson-distributed jumps observed from both the index and its component stocks constitute non-diversifiable risk.

We follow a framework similar to that of earlier works in jump risks in the CDS market. We first identify whether jump risks are present individually in each of the component sector sub-indices that comprise the iTraxx Non-Financial index as well as in the Non-Financial index itself. If significant jump risk is identified in the iTraxx Non-financial index, which acts as a portfolio of CDS's of the whole Non-financial market, we may conclude that such jump risk is non-diversifiable.

11.7 CDS Data and Preliminary Analysis

The International Index Company provides both bid quote and ask quote for each iTraxx index series. The data used in this study consists of daily closing mid-quotes for six iTraxx indices, namely iTraxx Non-Financials, iTraxx TMT, iTraxx Industrials, iTraxx Energy, iTraxx Autos and iTraxx Consumers. iTraxx Non-Financials is the index for the other 5 sub-indices and each sub-index accounts for a Non-financial sector. All these indices are traded with 5-year or 10-year maturity and are denominated in Euro. Here we just focus on the 5-year maturity, which is the most popular in the CDS market. The time period covered is March 22, 2004 to October 3, 2006. The total number of mid-point CDS spread quotes in the panel is 3900, that is 6 iTraxx index series each with 650 time points. All these data have been made available by the International Index company.

This section also provides some descriptive statistics of these indices which gives readers a rough idea of the dynamic behavior of a CDS index. Figure 11.5 shows these iTraxx indices for the sample period and Table 11.3 gives the descriptive statistics. In order to find whether these time series are stationary or not, we perform both the Augment Dickey-Fuller test and the Phillips-Perron test on our data. For robustness consideration, we try different model specifications and different lags. The results are quite similar and for simplicity, here we just list the results for the unit root test without trend and with four lags.

The Unit Root test shows that all iTraxx indices are non-stationary at the 5% significance level while all daily log-return on iTraxx indices are stationary. Except for the Energy sector, the size of the CDS spread, in basis points, varies somewhat more over the time period than across the sectors. Figure 11.5 displays the log difference of the iTraxx series.

Hans (2006), using the similar dataset, (for the period form June 21, 2004 to April 18, 2005), finds the distribution of the CDS index spread changes to be much more skewed and leptokurtic than the stock index return distribution. The iTraxx index is also at least two to three times as volatile (with a standard deviation equal to 30–40% on an annual basis) as the corresponding stock portfolio (with a standard deviation equal to 10–15% on an annual basis). Such a finding may also give us some hint about jump risk in the CDS market.

11.8 Methodology for Analyzing CDS Jump Risks

11.8.1 Normality Test for CDS Index Distribution

A non-parametric investigation quantifies what the visual inspection suggests, that neither levels nor log-return are normally distributed. Table 11.3 also lists excess kurtosis and skewness for the iTraxx index level and its daily log-return respectively. The Jarque-Bera test rejects the null hypothesis of normality for all iTraxx indices in our sample (p-values do not exceed even 0.01%). All CDS indices and their log-returns exhibit significant excess kurtosis, indicating that a pure diffusion most likely did not generate the data. Also, positive skewness for the iTraxx levels together with the excess kurtosis values confirms that large jumps are mostly positive.

11.8.2 Model for Individual iTraxx Indices

Similar to researchers in the stock market, we model the iTraxx index, which is a proxy for credit risk price, denoted by S_t and suppose this price evolves as follows:

$$\frac{dS_t}{S_t} = \gamma dt + \sigma dW_t + (\tilde{Y} - 1)dq_t \qquad (11.19)$$

where dW_t is a standard Brownian motion, dq_t represents a jump process and dW_t and dq_t are assumed to be orthogonal. dq_t is the jump term such that,

$$\begin{cases} 0 & \text{if no jump occurs} \\ 1 & \text{if jump occurs} \end{cases}. \qquad (11.20)$$

The jump process is assumed to be Poisson process with intensity λ, and $\tilde{Y}-1$ is the percentage changes in the asset price if the Poisson event occurs. $\tilde{Y}$ is assumed to have log normal distribution with mean θ and variance υ^2.

The sample path for S_t will be continuous most of the time with finite occasional jumps of different sizes and signs occurring at discrete points of time. If a jump occurs within a small time interval $(t, t+dt)$, the above stochastic process is

$$\frac{dS_t}{S_t} = \gamma dt + \sigma dW_t + \tilde{Y} - 1. \qquad (11.21)$$

If a jump does not occur within that time interval, the stochastic process is

$$\frac{dS_t}{S_t} = \gamma dt + \sigma dW_t. \qquad (11.22)$$

By Ito's lemma, the jump-diffusion process (11.19) can be expressed as

$$\frac{S_{t+dt}}{S_t} = e^{(\gamma-\frac{\sigma^2}{2})dt+\sigma dW_t} \tilde{Y}(J) = e^{\mu dt+\sigma dW_t}.\tilde{Y}(J) \qquad (11.23)$$

where J is the number of occurrences of Poisson distributed events during the interval of dt, which is characterized by single parameter λdt; $\tilde{Y}(J)=1$ if $J=0$, and $\tilde{Y}(J)=\prod_{j=1}^{J}\tilde{Y}_j$, if $J \geq 1$. Its discrete time version is

$$\frac{S_{t+\Delta t}}{S_t} = e^{\mu \Delta t+\sigma z_t} \tilde{Y}(J). \qquad (11.24)$$

By taking natural logarithm of both sides of (11.24), we obtain,

$$\text{Ln}(\frac{S_{t+\Delta t}}{S_t}) = \mu \Delta t + \sigma Z_t + \ln(\tilde{Y}(J)) \qquad (11.25)$$

where Z_t follows normal distribution with mean 0 and variance Δt. We can thus write,

$$\text{Ln}(\frac{S_{t+\Delta t}}{S_t}) \sim \sum_{j=0}^{J} \frac{e^{-\lambda \Delta t}(\lambda \Delta t)^j}{j!} N(\mu \Delta t + j\theta, \sigma^2 \Delta t + jv^2). \quad (11.26)$$

In this chapter the left hand side of (11.26) is treated as the return on iTraxx index.

The log-likelihood function is,

$$\text{LnL}(R_{iTraxx,t} : \mu, \sigma^2, \lambda, \theta, v^2) =$$

$$\sum_{t=1}^{T} \text{Ln}\left(\sum_{j=0}^{\infty} \frac{e^{-\lambda \Delta t}(\lambda \Delta t)^j}{j!} \times (2\pi(\sigma^2 \Delta t + jv^2))^{-0.5} \right.$$

$$\left. \times \exp\left[-\frac{(R_{itraxx,t} - (\mu \Delta t + j \cdot \theta))^2}{2(\sigma^2 \Delta t + j \cdot v^2)} \right] \right) \quad (11.27)$$

Basically, J can be any positive number, but for simplicity and easy computation workload, we set J equal to 1. The reason is that, the sample path for iTraxx index will be continuous most of the time with finite occasional jumps of different sizes and signs. Based on daily data we use, there is a small chance that CDS rate may jump twice within a day. The estimation results are listed in Table 11.4.

The Table 11.4 covers both the sector iTraxx indices and the iTraxx Non-Financials 5Y (which acts as an index for the sector iTraxx indices). It is apparent that all these indices experienced jumps during the sample period. The null hypothesis test that there is no jump ($\lambda = 0$) in the iTraxx sector index and iTraxx Non-financials based on the underlying stochastic process is rejected at 1% level.

Since the iTraxx Non-financial 5Y acts as a portfolio of all CDS's in Non-financial sector and it should have the benefit from diversification, but the results in Table 11.4 still reject the hypothesis $\lambda = 0$ even at the 1% significance level. Since the diversification does not remove the jump risk, then they may be driven by factor(s) other than firm or industry specific abnormal information. In other words, the results in Table 11.4 show that the jumps observed in iTraxx index may be systematic or non-diversifiable risk. From Table 11.4 we can also find that the estimated log-normally distributed mean jump sizes θ are all positive, which may provide some evidence that most jumps happening in the CDS market are positive, although they are mostly insignificant.

Comparing the variance of jump sizes and the instantaneous volatility conditional on no jumps confirms that the jump risk matters and the average variance of the jump size is around 3.5 times the instantaneous variance. Because the mean jump sizes are almost insignificant, for simplicity, in later estimation, we can impose the restriction that θ equals to 0 for computational ease.

11.8.3 Multivariate Analysis of Jumps in iTraxx Index with One Latent Common Factor

Here we adopt a vector autoregression type framework with common trend (Kim et. al (1999)) to model the dynamic behavior of the CDS spreads. The common factor is assumed to behave as a random walk with a deterministic trend.

Let ΔX_t denote daily log-return on each of the sector iTraxx indices, namely, iTraxx TMT 5Y, iTraxx Industrials 5Y, iTraxx Energy 5Y, iTraxx Autos 5Y and iTraxx Consumers 5Y. Let G_t be an unobserved single index which correspond to the log of the iTraxx Non-Financials 5Y. Thus ΔG_t will have an interpretation of being the log-return of the estimated common driver of the CDS indices. Mathematically, the model can be written as:

$$\Delta X_t = \mu_X + \beta \Delta G_t + v_t, v_t \sim N(0,\Sigma). \qquad (11.28)$$

Where, for simplicity, Σ is restricted to be a diagonal matrix,

$$\Sigma = \begin{bmatrix} \sigma_1^2 & 0 & 0 & 0 & 0 \\ 0 & \sigma_2^2 & 0 & 0 & 0 \\ 0 & 0 & \sigma_3^2 & 0 & 0 \\ 0 & 0 & 0 & \sigma_4^2 & 0 \\ 0 & 0 & 0 & 0 & \sigma_5^2 \end{bmatrix}. \qquad (11.29)$$

We also let,

$$\Delta G_t = \mu_G + \varepsilon_t, \varepsilon_t \sim N(0,\sigma_G^2). \qquad (11.30)$$

The model is in classical state space form and may be estimated using a Kalman filter framework. To clearly identify the measurement and state equation of the state space form we re-state the model in matrix difference notation. The measurement equation is given by:

$$
\begin{bmatrix} \Delta X_{1t} \\ \Delta X_{2t} \\ \Delta X_{3t} \\ \Delta X_{4t} \\ \Delta X_{5t} \end{bmatrix} = \begin{bmatrix} \mu_1 \\ \mu_2 \\ \mu_3 \\ \mu_4 \\ \mu_5 \end{bmatrix} + \begin{bmatrix} \beta_1 & 1 & 0 & 0 & 0 & 0 & 0 \\ \beta_2 & 0 & 1 & 0 & 0 & 0 & 0 \\ \beta_3 & 0 & 0 & 1 & 0 & 0 & 0 \\ \beta_4 & 0 & 0 & 0 & 1 & 0 & 0 \\ \beta_5 & 0 & 0 & 0 & 0 & 1 & 0 \end{bmatrix} \times \begin{bmatrix} \Delta G_t \\ v_{1,t} \\ v_{2,t} \\ v_{3,t} \\ v_{4,t} \\ v_{5,t} \\ G_{t-1} \end{bmatrix}, \quad (11.31)
$$

or,

$$
\Delta X_t = \mu_X + Z_t \alpha_t . \quad (11.32)
$$

The state vector α_t has the following dynamic form:

$$
\begin{bmatrix} \Delta G_t \\ v_{1,t} \\ v_{2,t} \\ v_{3,t} \\ v_{4,t} \\ v_{5,t} \\ G_{t-1} \end{bmatrix} = \begin{bmatrix} \mu_G \\ 0 \\ 0 \\ 0 \\ 0 \\ 0 \\ 0 \end{bmatrix} + \begin{bmatrix} 0 & 0 & 0 & 0 & 0 & 0 & 0 \\ 0 & 0 & 0 & 0 & 0 & 0 & 0 \\ 0 & 0 & 0 & 0 & 0 & 0 & 0 \\ 0 & 0 & 0 & 0 & 0 & 0 & 0 \\ 0 & 0 & 0 & 0 & 0 & 0 & 0 \\ 0 & 0 & 0 & 0 & 0 & 0 & 0 \\ 1 & 0 & 0 & 0 & 0 & 0 & 1 \end{bmatrix} \times \begin{bmatrix} \Delta G_{t-1} \\ v_{1,t-1} \\ v_{2,t-1} \\ v_{3,t-1} \\ v_{4,t-1} \\ v_{5,t-1} \\ G_{t-2} \end{bmatrix}
$$

$$
+ \begin{bmatrix} 1 & 0 & 0 & 0 & 0 & 0 & 0 \\ 0 & 1 & 0 & 0 & 0 & 0 & 0 \\ 0 & 0 & 1 & 0 & 0 & 0 & 0 \\ 0 & 0 & 0 & 1 & 0 & 0 & 0 \\ 0 & 0 & 0 & 0 & 1 & 0 & 0 \\ 0 & 0 & 0 & 0 & 0 & 1 & 0 \\ 0 & 0 & 0 & 0 & 0 & 0 & 0 \end{bmatrix} \times \begin{bmatrix} \varsigma_t \\ v_{1,t} \\ v_{2,t} \\ v_{3,t} \\ v_{4,t} \\ v_{5,t} \end{bmatrix} \quad (11.33)
$$

or,

$$
\alpha_t = \mu_\alpha + T_t \alpha_{t-1} + R_t \eta_t , \quad (11.34)
$$

where,

$$E(\eta_t \eta_t') = Q_t = \text{diag}\{\sigma_G^2, \sigma_1^2, \sigma_2^2, \sigma_3^2, \sigma_4^2, \sigma_5^2\}. \qquad (11.35)$$

The parameter set is given by the vector,

$$\Theta \equiv \{\mu_1, \mu_2, \mu_3, \mu_4, \mu_5, \mu_G, \beta_1, \beta_2, \beta_3, \beta_4, \beta_5, \sigma_1^2, \sigma_2^2, \sigma_3^2, \sigma_4^2, \sigma_5^2\}. \qquad (11.36)$$

The variance of the unobservable common component is set equal to 1, since the single factor in the model is assumed as a market portfolio. The estimation of such a state space model via Kalman filter has been described in Chapter. 1.

If we now suppose that the jump risk can not be diversified, that is the unobservable common factor may include a jump component in its sample path, so its process can be modified to include Poisson jumps as the problem now becomes multivariate analysis with one common factor including a jump component:

$$\Delta G_t = \mu_G + \varepsilon_t + \bar{y}, \varepsilon_t \sim N(0, \sigma_G^2) \qquad (11.37)$$

where, $\bar{y} \sim N(J \cdot \theta, J \cdot \sigma_Y^2)$ for $J = 0, 1$.

The probability density function (PDF) of innovation can be expressed as:

$$L = \sum_{j=0}^{\infty} \frac{\exp\{-\lambda\}\lambda^j}{j!} (2\pi(\sigma_G^2 + j \cdot \sigma_Y^2))^{-1/2} \\ \times \exp\left[-\frac{(\Delta G_t - \mu_G - j \cdot \theta)^2}{2(\sigma_G^2 + j \cdot \sigma_Y^2)} \right]. \qquad (11.38)$$

This type of model estimation has already been discussed in section 11.3. So the parameter set Θ is given by the vector,

$$\{\mu_1, \mu_2, \mu_3, \mu_4, \mu_5, \mu_G, \beta_1, \beta_2, \beta_3, \beta_4, \beta_5, \sigma_1^2, \sigma_2^2, \sigma_3^2, \sigma_4^2, \sigma_5^2, \lambda, \sigma_Y^2\}. \qquad (11.39)$$

There are eighteen parameters that need to be estimated and it is not easy for optimization. Theoretically, a different initial value should have no impact on the final optimal result for parameter estimation, however in practice it does.

We use a numerical optimization algorithm in Matlab 7.4 to estimate the parameters. In order to get stable results we use 20 sets of initial values for estimation work and choose the result with the largest value of log likelihood function. Although the estimators are slightly different based on their different upper and lower bounds (we keep the bound to a small value for lower computation burden), that does not change the key conclusion in this analysis that there is non-diversifiable positive jump in the CDS market.

11.9 Analysis of Results from the CDS Market

Table 11.5 lists the results for the ML estimation of one-factor model with jump component. It shows that the jump risk embedded in iTraxx sector indices is not diversifiable. The variance of jump sizes is much larger than the instantaneous variance conditional on no jumps, namely 36 versus 2. This result confirms again that the jump risk embedded in the CDS index is significant and undiversified.

The estimated value of λ, which is the mean rate of jumps per unit time is 0.1412 per day and the asymptotic standard error is 0.0283. Figure 11.7 gives the reader a visual impression of how the common factor of the CDS indices evolves over the sample period. Figure 11.7 also shows different suitable proxies of the common factor behind the entire sector index. Panel (a) gives the average log-sector index, which is calculated as:

$$\text{EWAIndex}_t = \frac{1}{5}(\log(\text{TMT}_t) + \log(\text{Industrial}_t)$$

$$+ \log(\text{Energy}_t) + \log(\text{Autos}_t) + \log(\text{consumers}_t)) . \quad (11.40)$$

Here TMT_t, Industrial_t Energy_t, Autos_t Consumers_t denote iTraxx TMT 5Y, iTraxx Industrials 5Y, iTraxx Energy 5Y, iTraxx Autos 5Y and iTraxx Consumers 5Y respectively.

Figure 11.7, panel (b) shows the one-period ahead predicted common factor by the one-factor model. From the plot we can see the model predicts the pattern of the common factor which underlies the entire sector indices quite well. Figure 11.7, panel (c) shows the iTraxx Non-Financials index along the same time axes. Compared with panel (c), time series plots in panel (a) and panel (b) are smoother.

From the plot, the whole CDS market experienced a significant rise from the end of April, 2005, reaching the peak around 20 May, 2005 and then returned to its average level. This is mainly affected by downgrade in Ford and GM debt that finally equated it with junk status by the ratings agency Standard & Poor's. Ford and General Motors are two of the world's biggest car companies, as recorded by the web site of www.financialpolicy.org, GM had $290 billion in outstanding debt and Ford had another $160 for a total of $450 billion. Given the massive size of the auto maker's debt, the downgrade had a sizeable impact on the whole market. The iTraxx Non-financial 5Y raised form 49.63 basis points on 5 May, 2005 to 66.04 basis points on 18 May, 2005. While the iTraxx Autos 5Y was up 16.45 basis points from 65.655 to 82.105 at the same period.

Table 11.6 gives correlation coefficient for these three series. Not surprisingly, these three series are highly correlated with each other.

Based on the estimated results, we conclude in this analysis, that the jump risks are present in iTraxx sector indices and these are not diversifiable. For robustness consideration, we also use a two-factor model (detailed results not reported here) to capture the characteristic of the dynamic behavior of iTraxx sector index. We find a similar result for the jump risk.

11.10 Summarizing CDS Market Application

The credit derivatives market has been developing rapidly over the last several years and among the different products, credit default swap (CDS) is the most popular type. This chapter examines this fast developing market and investigates the jump-diffusion process for iTraxx (a popular index for the CDS market) in a multivariate structural time-series setting for the stochastic process. Our empirical study detects significant Poisson-distributed jumps both in Non-financial iTraxx and its sub-indices.

The result that there is non-diversifiable jump risk in the CDS market found in this study is quite strong and robust based on the statistical analysis used in the analysis. Of course, further analysis together with alternative methodology and different dataset needs to be carried out. But the issue of non-diversifiable jump risk certainly warrants further empirical investigation.

Table 11.1 Maximum Likelihood Estimates of the Jump Diffusion Model.

	2002-2003	2004-2006
β_0	11.884 (3.432)	1.024 (0.007)
β_1	-7.196 (1.475)	0.010 (0.008)
β_2	0.910 (0.026)	1.010 (0.005)
β_3	1.035 (0.015)	1.010 (0.005)
β_4	0.816 (0.042)	1.011 (0.005)
β_5	1.061 (0.016)	1.011 (0.005)
β_6	1.300 (1.517)	1.045 (0.012)
α_0	0.119 (0.056)	1.052 (0.006)
α_1	0.555 (0.008)	0.340 (0.001)
α_2	0.259 (0.004)	0.339 (0.001)
μ_J	11.126 (1.214)	16.189 (0.956)
σ_J	240.360 (13.092)	1333.919 (57.094)
γ_1	1.971 (0.157)	2.145 (0.132)
γ_2	1.923 (0.146)	1.880 (0.117)
γ_3	2.484 (0.169)	1.943 (0.031)
γ_4	1.845 (0.144)	2.071 (0.038)
ϕ	0.164 (0.027)	0.082 (0.022)
TIC	0.3839	0.5489

Standard errors are in parentheses next to the parameters. TIC represents Theil's inequality coefficient, a measure of model's ability to predict the observed data. This is described in the text.

Table 11.2 iTraxx Europe Overview (Source: www.indexco.com).

Benchmark Indices	Sector Indices	Derivatives
iTraxx Europe	Non-Financials	Trenched iTraxx
125 investment grade entities	100 entities	Five standardized trenches of iTraxx Europe 0-3%, 3-6%, 6-9%, 9-12%, 12-22%
iTraxx Europe HiVol	Financials Sub	iTraxx Options
30 highest spread entities	Same 25 entities	
from iTraxx Europe	Financials Senior	
iTraxx Europe Crossover		iTraxx Futures
50 sub-investment grade		iTraxx Europe exposure traded
entities		as a futures contract

Table 11.3 Descriptive Statistics for iTraxx Index Level and Its Daily Log-Return.

	Non Fin.	TMT	Industrial	Energy	Auto	Consumers
Panel (a): Descriptive Statistics for iTraxx index						
Min	31.650	27.400	27.335	17.330	32.200	29.990
Mean	41.734	46.468	42.674	25.048	45.846	46.760
Media	41.103	45.305	43.695	24.050	44.683	47.053
Max	66.035	66.679	68.210	34.665	87.175	75.180
Stdev	5.298	8.186	6.089	3.763	8.000	8.531
Skewness	0.859	0.128	0.166	0.572	1.811	0.046
Kurtosis	4.892	2.478	4.548	2.340	7.944	2.989
ADF Test	0.189	0.236	0.098	0.113	0.164	0.571
PP test	0.163	0.315	0.134	0.172	0.097	0.602
Panel (b): Descriptive Statistics for daily log-return on iTraxx index						
Min$*10^2$	-10.522	-9.135	-9.221	-7.522	-15.810	-9.737
Mean$*10^2$	-0.061	-0.025	-0.031	-0.093	-0.041	-0.078
Median$*10^2$	-0.103	-0.089	-0.069	-0.042	-0.090	-0.100
Max$*10^2$	11.925	18.930	26.092	17.505	18.979	16.149
Stdev$*10^2$	1.989	2.162	2.169	2.031	2.404	1.930
Skewness	0.967	1.648	3.320	2.870	1.680	1.632
Kurtosis	11.531	17.309	38.882	23.820	22.067	17.246
ADF Test	0.000	0.000	0.000	0.000	0.000	0.000
PP Test	0.000	0.000	0.000	0.000	0.000	0.000

MacKinnon approximate p-value for Z (t) is listed in row ADF Test and PP Test. Non Fin is Non financial index.

Table 11.4 ML Estimates of the Jump-Diffusion Process (5-Year CDS).

	μ	σ^2	λ	θ	v^2	t
TMT	-0.115	1.124	0.169	0.534	3.485	5.62
	(0.057)	(0.045)	(0.030)	(0.527)	(0.279)	
Industrials	-0.148	0.978	0.176	0.669	3.696	6.65
	(0.050)	(0.037)	(0.027)	(0.560)	(0.187)	
Energy	-0.199	1.038	0.108	0.979	4.286	4.84
	(0.049)	(0.036)	(0.022)	(0.876)	(0.542)	
Autos	-0.163	1.026	0.145	0.836	4.709	6.50
	(0.049)	(0.036)	(0.022)	(0.639)	(0.362)	
Consumers	-0.162	1.165	0.106	0.792	3.638	4.33
	(0.055)	(0.042)	(0.025)	(0.680)	(0.447)	
Non-Financials	-0.162	1.041	0.154	0.651	3.352	5.46
	(0.051)	(0.041)	(0.028)	(0.483)	(0.392)	

Asymptotic standard error is reported below each estimated parameter and asymptotic t-statistic of is reported in the last column. Here we adopt BHHH estimator as variance estimators for an MLE. More detailed information may be found in William H. Greene, 'Econometric Analysis' fifth edition, P480-P482.

Table 11.5 ML Estimates of the one- factor Jump-Diffusion model (5-Year CDS).

	μ_X	β	σ^2
TMT	0.168	0.647	2.126
	(137.630)	(0.050)	(0.085)
Industrials	0.192	0.747	1.311
	(158.890)	(0.056)	(0.063)
Energy	0.075	0.561	2.211
	(119.31)	(0.045)	(0.077)
Autos	0.160	0.6742	3.011
	(143.335)	(0.054)	(0.060)
Consumers	0.0741	0.511	2.138
	(108.536)	(0.041)	(0.047)
Common Factor			
λ	0.1412		
	(0.028)		
μ_G	-0.439		
	(212.58)		
σ_Y^2	35.995		
	(7.181)		

$$\Delta X_t = \mu_X + \beta \Delta G_t + v_t, v_t \sim N(0,\Sigma)$$
$$\Delta G_t = \mu_G + \varepsilon_t + \overline{y}, \varepsilon_t \sim N(0,\sigma_G^2), \overline{y} \sim N(J \cdot \theta, J \cdot \sigma_Y^2) \text{ for } J = 0,1$$

X_t and G_t, respectively, denote daily log price of iTraxx index and an unobserved single common factor which is corresponding to the log of iTraxx Non-Financials 5Y. Asymptotic standard error for each parameter is also reported below each estimated parameter.

Table 11.6 Correlation between Equally-Weighted Average log iTraxx Sector Index, One Period Ahead Predicted Log Common Factor and Log iTraxx Non-Financials.

	Equally-weighted average	Predicted log common factor	Log iTraxx Non-financials
Equally-weighted average	1		
Predicted log common factor	0.962	1	
Log iTraxx Non-financials	0.970	0.913	1

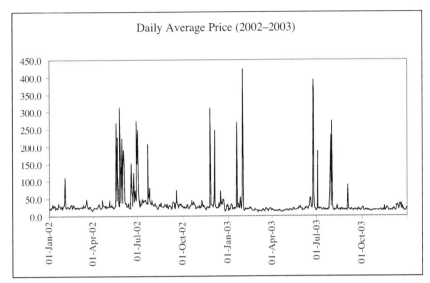

Figure 11.1 Daily Average Spot Electricity Prices for N.S.W. 2002–2003.

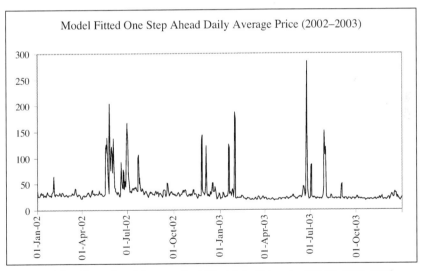

Figure 11.2 One Step Ahead Average Spot Electricity Prices for N.S.W. 2002–2003.

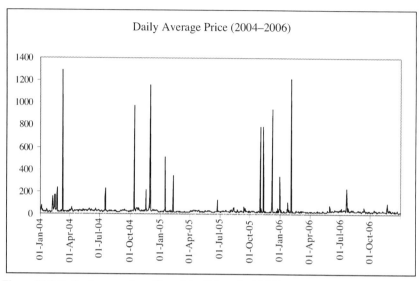

Figure 11.3 Daily Average Spot Electricity Prices for N.S.W. 2004–2006.

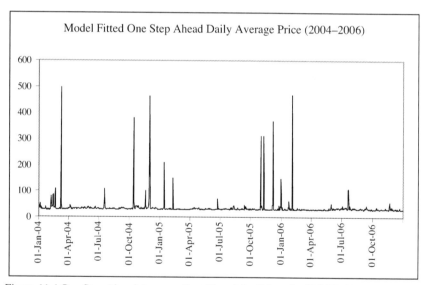

Figure 11.4 One Step Ahead Average Spot Electricity Prices for N.S.W. 2004–2006.

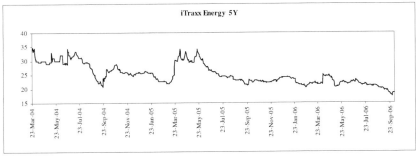

Figure 11.5 iTraxx Non-Financials 5Y and Its Sub-Index.

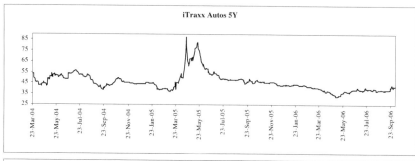

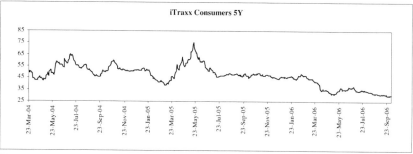

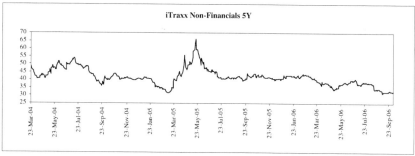

Figure 11.5 iTraxx Non-Financials 5Y and Its Sub-Index (Contd.).

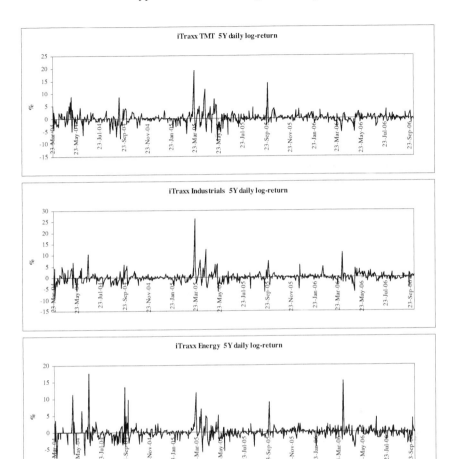

Figure 11.6 Daily Log-Return on iTraxx Non-Financials and Its Sub-Index.

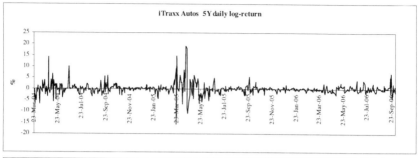

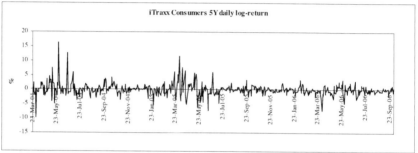

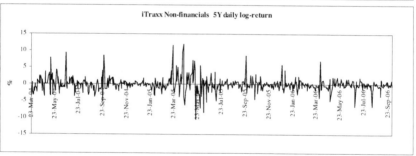

Figure 11.6 Daily Log-Return on iTraxx Non-Financials and Its Sub-Index (Contd.).

Panel (a) Average log-sector index (Equally weighted)

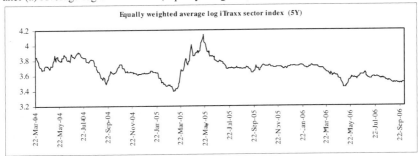

Panel (b) One-period ahead predicted common factor

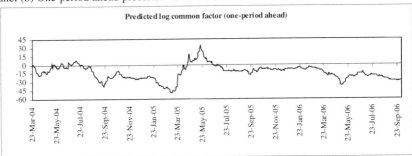

Panel (c)

Figure 11.7 Different Proxies for Common Factor in iTraxx Sector Index.

Bibliography

Abid, F. and Naifar, N. (2005), The impact of stock returns volatility on credit default swap rates: a copula study, International Journal of Theoretical & Applied Finance, 8, pp. 1135-1155.

Abid, F. and Naifar, N. (2006a), Credit-default swap rates and equity volatility: a nonlinear relationship, The Journal of Risk Finance, 7 (4), pp. 348-371.

Abid, F. and Naifar, N. (2006b), The determinants of credit default swap rates: an explanatory study, International Journal of Theoretical & Applied Finance, 9 (1), pp. 23-42.

Adjaoute, K. and Danthine, J-P. (2004), Portfolio diversification: Alive and well in Euroland, Applied Financial Economics, 14, pp. 1225-1231.

Ait-Sahalia, Y. (1996a), Nonparametric pricing of interest rate derivative securities, Econometrica, 64 (3), pp. 527-560.

Ait-Sahalia, Y. (1996b), Testing Continuous-Time Models of the Spot Interest Rate, The Review of Financial Studies, 9 (2), pp. 385-426.

Alois, L. J. G. and Stefan, P. (1999), A state-space approach to estimate and test multifactor Cox-Ingersoll-Ross models of the term structure, The Journal of Financial Research, 22, pp. 107-130.

Amato, J. D., and Remolona, E. M. (2004), The Credit Spread Puzzle, BIS Quarterly Review, 5, pp. 51-63.

Amin, K. I. and Morton, A. J. (1994), Implied volatility functions in arbitrage-free term structure models, Journal of Financial Economics, 35, pp.141-80.

Anderson, B. and Moore, J. B. (1979), Optimal Filtering, Englewood Cliffs, N.J., Prentice-Hall.

Avramov, D., Chordia, T., Jostova, G. and Philipov, A. (2007), Momentum and Credit Rating, The Journal of Finance, 62 (5), pp. 2503-2520.

Babbs S. H., Nowman K. B. (1999), Kalman Filtering of Generalized Vasicek Term Structure Models" Journal of Financial and Quantitative Analysis, 34 (1), pp. 118-130.

Backus, D., Gregory, A. and Telmer, C. (1993), Accounting for forward rates in markets for foreign currency, Journal of Finance, 48, 1887-1908.

Bakshi, G.S. and Naka, A. (1997), Unbiasedness of the forward exchange rates, The Financial Review, 32(1), 145-162.

Bakshi, G. and Kapadia, N. (2003), Delta-Hedged Gains and the Negative Market Volatility Risk Premium, The Review of Financial Studies, 16, pp. 527-566.

Bali, T. G. and Weinbaum, D. (2007), A conditional extreme value volatility estimator based on high-frequency returns, Journal of Economic Dynamics and Control, 31 (2), pp. 361-397.

Ball, C. A., Torous, W. N., (1999), The Stochastic Volatility of Short-Term Interest Rates: Some International Evidence, The Journal of Finance, 54 (6), pp. 2339-2359.

Ball, L. and Tchaidze, R. (2002), The Fed and the new economy, American Economic Review, Papers and Proceedings, 92 (2), pp. 108-114.

Barro, R, and Sala-i-Martin X. (1995), Economic growth, McGraw-Hill, New York.

Barro, R. J. (1991), Economic growth in a cross-section of counties, Quarterly Journal of Economics, 106, pp. 407–443.

Barro, R. J. and Sala-i-Martin, X. (1994), Convergence, Journal of Political Economy, 100, pp. 223–251.

Bekaert, G. (1994), Exchange rate volatility and deviation from unbiasedness in a cash-in-advance model, Journal of International Economics, 36, 29-52.

Benkert, C. (2004), Explaining credit default swap premia, The Journal of Futures Markets, 24 (1), pp.71-92.

Bernard, A. B. (1992), Empirical Implications of the Convergence Hypothesis, Working Paper, MIT, Cambridge, MA.

Bernard, A. B. and Durlauf, S. N. (1995), Convergence in international output, Journal of Applied Econometrics, 10, pp. 97–108.

Berument, H. and Jelassi, M. M. (2002), The Fisher hypothesis: A multi-country analysis, Applied Economics, 34 (13), pp. 1645-1655.

Berument, H., Kilinc, Z. and Ozale, U. (2005), The missing link between inflation uncertainty and interest rates, Scottish Journal of Political Economy, 52 (2), pp. 222-241.

Beveridge, S. and Nelson, C. R., (1981), A new approach to decomposition of economic time series into permanent and transitory components with particular attention to measurement of the business cycle, Journal of Monetary Economics, 7, pp. 151-174.

Bhar, R. and Chiareiia, C. (1995a), The estimation of the Heath-Jarrow-Morton model by use of Kalman filtering techniques, Working Paper No. 54, School of Finance and Economics, University of Technology, Sydney, December 1995.

Bhar, R. and Chiarella, C. (1995b), Transformation of Heath-Jarrow-Morton Models to Markovian Systems, School of Finance & Economics, University of Technology, Sydney, Working paper No. 53, December, 1995.

Bhar, R. and Chiarella, C. (1997), The Estimation of Heath-Jarrow-Morton Model by Use of Kalman Filtering Techniques, Computational Approaches to Economic

Problems, eds. H. Amman et al, Kluwer Academic Publishers, The Netherlands, pp. 113-126.

Bhar, R. and Hamori, S. (2004), Hidden Markov Models: Application to Financial Economics, Springer, Heidelberg, July.

Bhar, R. and Hunt, B. F. (1993), Predicting the Short-term Forward Interest Rate Structure Using a Parsimonious Model, Review of Futures Markets, 12 (3), pp. 577-590.

Bhar, R., Chiarella, C. and Pham, T (2001), Modeling the Currency Forward Risk Premium: A New Perspective, Asia-Pacific Financial Markets, 8(4), 341-360.

Bhar, R., Chiarella, C. and Runggaldier, W. (2004), Inferring the forward-looking equity risk premium from derivative prices, Studies in Nonlinear Dynamics & Econometrics 8, Article 3.

Bishop, G., and Welch, G. (2001), An Introduction to the Kalman Filter, Lecture notes, University of North Carolina, Department of Computing Science.

Black, F. and Cox, J. C. (1976), Valuing corporate securities- some effects of bond indenture provisions, The Journal of Finance, 31, pp. 351-367.

Black, F. and Scholes, M. (1973), The pricing of options and corporate liabilities, The Journal of Political Economy, 81, pp. 637-654.

Blair, B. J., Poon Ser-Huang, and Taylor, S. J. (2001), Forecasting S&P 100 volatility: The incremental information content of implied volatilities and high-frequency index returns, Journal of Econometrics, 105, pp. 5-26.

Bleaney, M. F., Leybourne, S. J. and Mizen, P. (1999), Mean reversion of real exchange rates in high-inflation countries, Southern Economic Journal, 65 (4), pp. 839-854.

Boudoukh, J. and Richardson, M. (1993), Stock returns and inflation: a long horizon perspective, American Economic Review, 83, pp. 1346-1355.

Brenner, R.J., Harjes, R.H., and Kroner, K.F. (1996), Another Look at Models of the Short-Term Interest Rate, Journal of Financial and Quantitative Analysis, 31(1), pp. 85-107.

Brigo, D. and Mercurio, F. (2006), Interest rate models -Theory and practice (2nd Ed.), Springer, New York, Berlin, Heidelberg.

Broto, C., Ruiz, E., (2004), Estimation methods for stochastic volatility models: A survey, Journal of Economic Surveys, 18, pp. 613-649.

Brown, B., In, F., and Fang, V. (2002), Modeling the determinants of swap spreads, The Journal of Fixed Income, 12, pp.29-40.

Bruno, M. and Easterly, W. (1998), Inflation Crises and Long-run Growth, Journal of Monetary Economics, 41, pp. 3-26.

Bystrom, H. (2006), Credit Grades and the iTraxx CDS Index Market, Financial Analysts Journal, 62 (6), pp. 65-76.

Cameron, N., Hum, D., and Simpson, W., (1996), Stylized Fact and Stylized Illusions: Inflation and Productivity Revisited, Canadian Journal of Economics, 30, pp. 152-162.

Campbell, J. Y. and Shiller, R. J. (1988a), The Dividend-Price Ratio and Expectations of Future Dividends and Discount Factors, Review of Financial Studies, 1, pp. 195-228.

Campbell, J.Y. and Shiller, R. J. (1988b), Stock Prices, Earnings and Expected Dividends, Journal of Finance, 43, pp. 661-676.

Campbell, J.Y. and Kyle, A. S. (1993), Smart Money, Noise Trading and Stock Price Behavior, Review of Economic Studies, 61, pp. 1-34.

Campbell, M., Lettau, B.G., Malkiel, J., and Xu, Y. (2001), Have individual stocks become more volatile? Journal of Finance, 56, pp. 1-43.

Campbell, Y., and Taksler, G. B. (2003), Equity volatility and corporate bond yields, Journal of Finance, 58, pp. 2321-2349.

Carverhill, A. (1994), When is the short rate Markovian? Mathematical Finance, 4, pp. 305-312.

Cashin, P., McDermott, C. J. and Scott, A. (1999), The myth of co-moving commodity prices, Reserve Bank of New Zealand Discussion Paper G99/9, Wellington.

Cavaglia, S. (1992), The persistence of real interest differentials: A Kalman filtering approach, Journal of Monetary Economics, 29, pp. 429-443.

Chakravarti, I., Laha, R. G., and Roy, J. (1967), Handbook of Methods of Applied Statistics, Vol. I. (Wiley and Sons), pp. 392-394.

Chaleyat-Maurel, M. and Michel, D. (1984), Des resultants de no-existence de filter de dimension finite, Stochastics, 13, pp. 83-102.

Chan, K. C., Karolyi, G. A., Longstaff, F. and Sanders, A. (1992), The volatility of short-term interest rates: An empirical comparison of alternative models of term structure of interest rates, Journal of Finance, 47, pp. 1209-1227.

Chen, H. (2007), Macroeconomic conditions and the puzzles of credit spreads and capital structure, Working paper, University of Chicago.

Chen, L., Collin-Dufresne, P., and Goldstein, R. S. (2008), On the Relation between the Credit Spread Puzzle and the Equity Premium Puzzle, AFA Boston Meetings paper Available at SSRN.

Chen, R. R., and Scott, L. (1993), Multi-factor Cox-Ingersoll-Ross models of the term structure: Estimates and tests from a Kalman Filter model, Journal of Fixed Income 3, pp. 14-31.

Cheung, Y. (1993), Exchange rate risk premiums, Journal of International Money and Finance, 12, pp. 182-194.

Chiang R., Davidson, I. and Okunev, J. (1996), Some Further Theoretical and Empirical Implications Regarding the Relationship Between Earnings, Dividends and Stock Prices, Working paper 60, Faculty of Business, University of Technology, Sydney.

Chiareiia, C. and El-Hassan, N. (1996), The solution of partial differential equations arising in HJM theory of the term structure of interest rates. Presented at the 3rd Japanese Association of Financial Econometrics and Engineering International Conference on Investments and Derivatives, Tokyo, May 1996.

Chiarella, C. and El-Hassan, N. (1996a), A Preference Free Partial Differential Equation for the Term Structure of Interest Rates, Financial Engineering and the Japanese Market, 3(3), pp. 217-238.

Chiarella, C. and El-Hassan, N. (1996b), Evaluation of Derivative Security prices in the Heath-Jarrow-Morton Framework as Path Integrals Using Fast Fourier Transform Techniques, Paper presented at the Graduate School of Business, Stanford University, August 23.

Chiarella, C. and El-Hassan, N. (1997), A Tutorial on Yield Curve Analysis, School of Finance & Economics, University of Technology, Sydney, Working paper No. 75.

Chou, R., Engle, R. and Kane, A. (1992), Measuring risk aversion from excess returns on a stock index, Journal of Econometrics, 52, pp. 201-224.

Choudhry, T. (2002), The stochastic structure of the time varying Beta: evidence from UK companies, The Manchester School, 70, pp. 768-791.

Chow, Sy-Miin., Ferrer, E. and Nesselroade, J. R. (2007), An Unscented Kalman Filter approach to the estimation of nonlinear dynamical system model, Multivariate Behavioral Research, 42(2), pp. 283-321.

Christiansen, C. (2005), Multivariate term structure models with level and heteroskedasticity effects, Journal of Banking and Finance, 29, pp. 1037-1057.

Christensen, B. J. and Prabhala, N. R. (1998), The relation between implied and realized volatility, Journal of Financial Economics, 50, pp. 125-150.

Christopher, L. C. and Andrea, M. P. N. (1998), Credit and interest rate risk in the business of banking, Derivatives Quarterly, 4, pp.19-35.

Chunsheng, Z. (2001), The term structure of credit spreads with jump risk, Journal of Banking & Finance, 25, pp. 2015-2040.

Clark, T. E. (1997), Cross-country evidence on long-run growth and inflation, Economic Inquiry, 35, pp. 70-81.

Cochrane, John H. (1988), How big is the random walk in GNP?, Journal of Political Economy, 91, pp. 893-920.

Collin-Dufresne, P. and Goldstein, R. S. (2001), Do credit spreads reflect stationary leverage ratios?. The Journal of Finance, 56 (5), pp. 1929-1957.

Collin-Dufresne, P., Goldstein, R. S. and Martin, J. S. (2001), The determinants of credit spread changes, The Journal of Finance, 56 (6), pp. 2177-2207.

Collin-Dufresne, P., Goldstein, R. and Hugonnier, J. (2004), A general formula for valuing defaultable securities, Econometrica, 72 (5), pp. 1377-1407.

Collin-Dufresne, R., Goldstein, P., and Helwege, J. (2003), Is Credit Event Risk Priced? Modeling Contagion via the Updating of Beliefs, Working paper.

Cont, R. (2001), Empirical properties of asset returns: stylized facts and statistical issues, Quantitative Finance, - , pp. 223–236.

Coval, J. D. and Shumway, T. (2001), Expected option returns, The Journal of Finance, 56, pp. 983-1009.

Cox J., Ingersoll, J., and Ross, S. (1985), A theory of the term structure of interest rates, Econometrica, 53, pp. 385-407.

Cox, J. C., Ingersoll, J. E. and Ross, S. A. (1981), The Relation Between Forward Prices and Futures Prices, Journal of Financial Economics, 9, pp. 321-346.

Davidson, I., Okunev, J. and. Tahir, M. (1996), Modeling the Equity Risk Premium in the Long Term, Working paper 59, Faculty of Business, University of Technology, Sydney.

Davis, M. H. A. and Marcus, S. I. (1981). An introduction to nonlinear filtering, in Stochastic Systems: The Mathematics of Filtering and Identification and Application, Reidel, Dordrecht, pp. 53-75.

Das, S. R., (2002), The surprise element; jumps in interest rates. Journal of Econometrics, 106, pp. 27-65.

De Jong, F., and Driessen, J. (2006), Liquidity Risk Premia in Corporate Bond Markets, Working paper, University of Amsterdam.

De Jong, P., Shephard, N., (1995), The simulation smoother for time series models, Biometrika, 82, pp. 339-350.

Delianedis, G., and Geske, R. (2002), The Components of Corporate Credit Spreads: Default, Recovery, Tax, Jumps, Liquidity, and Market Factors, Working paper 22-01, Anderson School, UCLA.

Demetriadas, P. (1988), Macroeconomic aspects of the correlation between the level and the variability of Inflation, Economic Letters, 26 (2), pp. 121-124.

Dowrick, S., Nguyen, D. T. (1989), OECD comparative economic growth 1950–1985: Catch-up and convergence, American Economic Review, 79, pp. 1010–1030.

Driessen, J. (2005), Is Default Event Risk Priced in Corporate Bonds? Review of Financial Studies, 18, pp. 165-195.

Duffee, D., and Singleton, K. (1999), Modelling the Term Structure of Defaultable Bonds, Review of Financial Studies, 12, pp. 687-720.

Duffee, G. (1993), On the relation between the level and volatility of short-term interest rates: A comment on Chan, Karolyi, Longstaff and Sanders, Working paper, Federal Reserve Board, Washington, D.C.

Duffee, G. R. (1998), The relation between treasury yields and corporate bond yield spreads, The Journal of Finance, 53 (6), pp. 2225-2241.

Duffee, G. R. and Zhou, C. S. (2001), Credit derivatives in banking: Useful tools for managing risk? Journal of Monetary Economics, 48 (1), pp. 25-54.

Duffee, G.R. (1998), The relation between treasury yields and corporate bond yield spreads, Journal of Finance, 53, pp. 225-2241.

Duffie, D. and Singleton, K. J. (1999), Modeling term structures of defaultable bonds, The Review of Financial Studies, 12, pp. 687-720.

Durlauf, S. Johnson P. (1992), Local vs. global convergence across national economies, NBER working paper no. 3996, Cambridge, Massachusetts.

Duffie, D., Pan, J. and Singleton, K. (2000), Transform Analysis and Asset Pricing for Affine Jump-Diffusions, Econometrica, 68 (6), pp. 1343-1376.

Durbin, J., Koopman, S. J. (1997), Monte Carlo maximum likelihood estimation of non-Gaussian state space model, Biometrika, 84 (3), pp. 669-684.

Durbin, J., Koopman, S. J. (2002), A simple and efficient simulation smoother for state space time series analysis, Biometrika, 89 (3), pp. 603-615.

Efron, B. (1987), Better Bootstrap Confidence Intervals, Journal of American Statistical Association, 82, pp. 171-200.

Elton, E. J., Agrawal, D., Gruber, M. J., and Mann, C. (2001), Explaining the Rate Spread on Corporate Bonds, Journal of Finance, 56, pp. 247-277.

Elyasiani, E., and Mansur, I. (1998), Sensitivity of the bank stock returns distribution to changes in the level and volatility of interest rate: A GARCH-M model, Journal of Banking and Finance, 22, pp. 535-563.

Engel, C. (1996), The forward discount anomaly and the risk premium: A survey of recent evidence, Journal of Empirical Finance, 3, 123-192.

Eom, Y. H., Helwege, J., and Huang, J. (2004), Structural Models of Corporate Bond Pricing: An Empirical Analysis, Review of Financial Studies, 17, pp. 499-544.

Ericsson, J. and Reneby, J. (2004), An empirical study of structural credit risk models using stock and bond prices, The Journal of Fixed Income, 13 (4), pp. 38-49.

Ericsson, J., and Renault, O. (2006), Liquidity and Credit risk, Journal of Finance, 61, pp. 2219-2250.

Ericsson. J, Jacobs, K and Oviedo, R. (2009), The Determinants of Credit Default Swap Premia, Journal of Financial and Quantitative Analysis, 44 (1), pp. 109-132.

Estrella, A., and Hardouvelis, G. A. (1991), The Term structure as a predictor of Real Economic Activity, Journal of Finance, 46, pp. 555-576.

Faff, R. W., Hillier, D. and Hillier, J. (2000), Time varying beta risk: an analysis of alternative modeling techniques, Journal of Business Finance & Accounting, 27, pp. 523-554.

Fahmy, R. and Kandil, M. (2002), The Fisher effect: New evidence and implications, International Review of Economics and Finance, 12 (4), pp. 451-465.

Falb, P. L. (1967), Infinite-dimensional filtering: the Kalman-Bucy filter in Hilbert space, Information and Control, 11, pp. 102-137.

Fama, E. F. and French, K. R. (1988), Permanent and temporary components of stock prices, Journal of Political Economy, 96, pp. 246-273.

Fama, E. F. and French, K. R. (1989), Business conditions and expected returns on stocks and bonds, Journal of Financial Economics, 25, pp. 23–49.

Fama, E. F., and French, K. R. (1996), Multifactor Explanations of Asset Pricing Anomalies, Journal of Finance, 51, pp. 55–84.

Feldhutter, P., and Lando, D. (2008), Decomposing Swap Spreads, Journal of Financial Economics, 88, pp. 375–405.

Ferson, W. and Harvey, C. (1991), The variation of economic risk premiums, Journal of Political Economy, 99, pp. 23-49.

Fisher, L. (1959), Determinants of risk premiums on corporate bonds, Journal of Political Economy, 67, pp. 217–237.

Flesaker, B. (1993), Arbitrage Free Pricing of Interest Rate Futures and Forward Contracts, The Journal of Futures Markets, 13 (1), pp. 77-91.

Flesaker, B. (1993), Testing the Heath-Jarrow-Morton/Ho-Lee model of interest rate contingent claims pricing, Journal of Financial and Quantitative Analysis, 28, pp. 483-485.

Fong, H. G. and Vasicek, O. A. (1992), Interest rate volatility as a stochastic factor, Working paper, Gilford Fong Associates.

Fountas, S., Karanasos, M. and Kim, J. (2002), Inflation and output growth uncertainty and their relationship with inflation and output growth, Economic letters, 75, pp. 293-301.

Francis, A. L., Sanjay, M. and Eric, N. (2005), Corporate Yield Spreads: Default Risk or Liquidity? New Evidence from the Credit Default Swap Market, The Journal of Finance, 60, pp. 2213-2253.

Frenkel, J. and Razin, A. (1986), Fiscal policies in the world economy, Journal of Political Economy, 94, pp. 564-594.

Fridson, M., and Jonsson, J. (1995), Spread versus treasuries and the riskiness of high-yield bonds, Journal of Fixed Income, 5, pp. 79–88.

Friedman, M. (1973), Money and Economic Development, Toronto, Leington Books.

Friedman, M. (1977), Nobel lecture: Inflation and Unemployment, Journal of Political Economy, 85 (3), pp. 452-472

Geman, H. and Roncoroni, A. (2006), Understanding the Fine Structure of Electricity Prices, Journal of Business, 79 (3), pp. 1225-1261.

Gemmill, G., and Keswani, A., (2008), Idiosyncratic Downside Risk and the Credit Spread Puzzle, Working paper, available at SSRN.

Geyer, A. L. J. and Pichler, S. (1999), A state-space approach to estimate and test multifactor Cox-Ingersoll-Ross models of the term structures, The Journal of Financial Research, 22 (1), pp. 107-130.

Gibbons, M. R. and Ferson, W. (1985), Testing Asset Pricing Models with Changing Expectations and an Unobservable Market Portfolio, Journal of Financial Economics, 14, pp. 217-236.

Gikhman, I. I. and Skorokhod, A. V. (1965), Introduction to the Theory of Random Processes, Philadelphia, PA: Saunders.

Granger, C. W. J. and Swanson, N. R. (1997), An introduction to stochastic unit root processes, Journal of Econometrics 80, pp. 35-62.

Grier, K. B. and Perry, M. (2000), The effects of real and nominal uncertainty on inflation and output growth: some GARCH-M evidence, Journal of Applied Econometrics, 15, pp. 45-58.

Grier, K. B., Henry, O. T., Olekalns, N. and Shields, K. (2004), The effects of real and nominal uncertainty on inflation and output growth: some GARCH-M evidence. Journal of Applied Econometrics, 15, pp. 45-58.

Grier, R. and Grier K. B. (2006), On the real effects of inflation and inflation uncertainty in Mexico, Journal of Development Economics, 80, pp. 478-500.

Groenewold, N. and Fraser, P. (1999), Time varying estimates of CAPM betas, Mathematics and Computers in Simulation, 48, pp. 531-539.

Gujarati (2004), Basic Econometrics, Fourth Edition, The McGraw-Hill Companies.

Hall S, and Aubyn M. St. (1995), Using the Kalman Filter to test for convergence: A comparison to other methods using artificial data. documento de trabalho 11/95, Departamento de Economia, Instituto Superior de Economia e GestaÄo, Lisbon.

Hall. S., Robertson, D. and Wickens, M. (1992), Measuring convergence of the EC economies. Manchester School of Economics and Social Studies, 60, Supplement: 99-111.

Hamilton, J. D., (1994), Time Series Analysis, Princeton University Press, Cambridge.

Hans, B. (2006), Credit Grades and the iTraxx CDS Index Market, Financial Analysts Journal, 62, pp. 65-76.

Harding, D. and Pagan, A. (1999), Dissecting the cycle, Melbourne Institute Working Paper No. 13/99, The University of Melbourne, Australia.

Harvey, A. C. (1989), Forecasting structural time series models and the Kalman filter, Cambridge University Press, Cambridge.

Harvey, A. C. (1990), Forecasting, Structural Time Series Models and the Kalman Filter, Cambridge: Cambridge University Press.

Harvey, A. C., (1990), The Econometric Analysis of Time Series, 2nd edition. Cambridge, Massachusetts: The MIT Press.

Harvey, A. C., Ruiz, E. and Sentana, E. (1992), Unobserved Component Time Series Models with ARCH Disturbances, Journal of Econometrics, 52, pp. 129-157.

Harvey, A., Ruiz, E., Shephard, N. (1994), Multivariate stochastic variance models, Review of Economic Studies, 61, pp. 247-264.

Harvey, A. C., Shephard, N. (1993), Estimation and testing of stochastic variance models. Mimeo, London School of Economics.

Hatgioannides, J., Karanasos, M. and Karanassou, M. (2004), Permanent and Transitory Components in a Continuous-Time Model of the Term Structure, WSEAS Transactions on Business and Economics, 1(2, April)

Hayford, M. D. (2000), Inflation uncertainty, unemployment uncertainty and economic activity, Journal of Macroeconomics. 22, pp. 315-329.

Heath, D., Jarrow, R. and Morton, A. (1990), Contingent claim valuation with a random evolution of interest rates, Review of Futures Markets, 9, pp. 54-82.

Heath, D., Jarrow, R. and Morton, A. (1992a), Bond pricing and the term structure of interest rates: a new methodology for contingent claims valuation, Econometrica, 60, pp. 77-105.

Heath, D., Jarrow, R. and Morton, A. (1992b), Easier done than said, RISK, 5(9), pp. 77-80.

Heston, S. L. (1993), A closed-form solution for options with stochastic volatility with applications to bonds and currency options, The Review of Financial Studies, 6(2), pp. 327–343.

Ho, T. S. Y. and Lee, S. B. (1986), Term structure movements and pricing interest rate contingent claims, Journal of Finance, 41, pp. 1011-1029.

Huang, J., and Huang, M. (2003), How much of the corporate-treasury yield spread is due to credit risk? Working paper, Pennsylvania State University.

Huang, J., and Kong, W. (2003), Explaining credit spread changes: New evidence from option-adjusted bond indexes, Journal of Derivatives, Fall, pp. 30-44.

Huizinga, J. (1993), Inflation uncertainty, relative price uncertainty, and investment in US manufacturing, Journal of Money, Credit and Banking, 25 (3), pp. 521-549.

Hull, J. and White, A. (1990), Pricing Interest Rate Derivative Securities, Review of Financial Studies, 23(3), pp. 573-592.

Hull, J. and White, A. (1994), Numerical Procedures of Implementing Term Structure models II: Two-Factor Models, The Journal of Derivatives,.2, pp. 37-47.

Hull, J. C. (1997), Options, Futures, and Other Derivatives, Third Edition, Prentice Hall International Inc.

Hull, J., Predescu, M. and White, A. (2004), The relationship between credit default swap spreads, bond yields, and credit rating announcements, Journal of Banking and Finance, 28, pp. 2789-2811.

Jacobs, K., and Li, X. (2008), Modeling the Dynamics of Credit Spreads with Stochastic Volatility, Management Science, 54, pp. 1176-1188.

Jacquier, E., Polson, N.G., Rossi, P.E., (1994), Bayesian analysis of stochastic volatility models, Journal of Business and Economic Statistics, 12, pp. 371-417.

Jan, E. and Olivier, R. (2006), Liquidity and Credit Risk, The Journal of Finance, 61, pp.2219-2250.

Jarrow, R. A. J. and Yildiray, Y. (2002), Valuing default swaps under market and credit risk correlation, The Journal of Fixed Income,.11, pp. 7-19.

Jarrow, R. A., and Turnbull, S. M. (1995), Pricing derivatives on financial securities subject to credit risk, Journal of Finance, 50, pp. 53-85.

Jarrow, R. A., Lando, D., and Yu, F. (2001), Default risk and diversification: theory and applications, Working paper, University of California-Irvine.

Jarrow, R. and Turnbull, S. M. (1994), Delta, Gamma and Bucket Hedging of Interest Rate Derivatives, Applied Mathematical Finance, 1, pp. 21-48.

Javaheri, A., Lauteir, D. and Galli, A. (2005), Filtering in Finance, Wilmott Magazine, www.wilmott.com.

Jayaraman, T. K., Ward, B. D., and Xu, Z. L. (2005), Are the Pacific Islands ready for a currency union? An empirical study of degree of economic convergence, USPEC Working Paper, No. 2005/2, February.

Jazwinski, A. H. (1970), Stochastic Processes and Filtering Theory, Academic Press, New York.

Jenrich, R.I. (1970), An asymptotic χ^2 test for the equality of two correlation matrices, Journal of the American Statistical Association, 65, pp. 904-912.

Jochum, C. (1999), Volatility spillovers and the price of risk: evidence from the Swiss stock market, Empirical Economics, 24, pp. 303-322.

Johnson, D. R. (2002), The effect of inflation targeting on the behavior of expected inflation: evidence from an 11 country panel. Journal of Monetary Economics 49, 1521-1538.

Jones, E. P., Mason, S. P. (1984), Contingent Claims Analysis of Corporate Capital Structures: An Empirical Investigation/Discussion, The Journal of Finance, 39, pp.611-627.

Jong, F and Santa-Clara, P. (1999), The dynamics of the forward interest rate curve: A formulation with state variables, Journal of Financial and Quantitative Analysis, 34, pp. 131-157.

Joost, D. (2005), Is Default Event Risk Priced in Corporate Bonds? The Review of Financial Studies, 18, pp.165-195.

Joutz, F., Mansi, S. A., and Maxwell, W. F. (2001), The Dynamics of Corporate Credit Spreads, Working paper, George Washington University and Texas Tech University.

Julier, S. J., Uhlmann, J. K and Durrant-Whyte, H. F. (1995), A new approach for filtering nonlinear systems, Proceedings of the American Control Conference, Seattle, Washington, IEEE. pp. 1628–1632

Julier, S. J. and Uhlmann, J. K. (1997), A New Extension of the Kalman Filter to Nonlinear Systems, The Proceedings of AeroSense: The 11th International Symposium on Aerospace/Defense Sensing, Simulation and Controls, Orlando, FL, USA, 1997.

Julier, S. J., and Uhlmann, J. K. (2002), Reduced sigma point filters for the propagation of means and covariance through nonlinear transformation, Proceedings of the IEEE American Control Conference Anchorage AK, USA, IEEE. pp. 887–892.

Kalman, R. E. (1960), A new approach to linear filtering and prediction problems, Transactions of the ASME–Journal of Basic Engineering, 82 (Series D), pp. 35–45.

Kandil, M. (2005), Money, interest, and prices: some international evidence, International Review of Economics and Finance, 14, pp. 129-147.

Kim, C-J., and Nelson, C. R. (1999), State Space Models with Regime Switching: Classical and Gibbs-Sampling Approaches with Applications, The MIT Press, Cambridge.

Kim, M-J., Oh, Y-H. and Brooks, R. (1994), Are Jumps in Stock Returns Diversifiable? Evidence and Implications for Option Pricing, Journal of Financial and Quantitative Analysis, 29 (4), pp. 609-631.

Kloeden, P. E. and Platen, E. (1992), The Numerical Solution of Stochastic Differential Equations, Springer-Verlag, Berlin.

Kloeden, P. E., Schurz, H., Platen, E. and Sorensen, M.(1992), On Effects of Discretisation on Estimators of Drift Parameters for Discussion Processes, Working paper No. 249, June, Department of Theoretical Statistics, University of Aarhus.

Kontonikas, A. (2004), Inflation and Inflation uncertainty in the United Kngdom, evidence from GARCH modeling. Economic Modeling 21 525-543.

Kwan, S. H. (1996), Firm-Specific Information and the Correlation Between Individual Stocks and Bonds, Journal of Financial Economics, 40, pp. 63-80.

Langetieg, T. C., (1980), A multivariate model of the term structure, Journal of Finance, 35, pp. 71-91.

Lautier D. (2000), La Structure par Terme des Prix des Commodités : Analyse Théorique et Applications au Marché Pétrolier, Thèse de Doctorat, Université Paris IX.

Lautier, D. and Galli, A. (2001), Un modèle des prix à termes des matières premières avec rendement d'opportunité asymétrique, FINECO, 11, pp.73-93

Levin, R. and Renelt, D. (1992), A sensitivity analysis of cross-country growth regressions, American Economic Review, 82, pp. 942-963.

Levin, R. and Zervos, S. (1993), What have we learned about policy and growth from cross-country analysis, American Economic Review Papers and Proceedings, 83, pp. 426-430.

Lim, L. K. and McAleer, M. (2003), Convergence and catching up in ASEAN: A comparative analysis, CIRJE Discussion Papers, available at http://www.c.u-tokyo.ac.jp/cirje/research/03research02dp.html .

Linter, J. (1956), Distribution of Incomes of Corporations Among Dividends, Retained Earnings, and Taxes, American Economic Review, 61, pp. 97-113.

Lipster, R. S. and Shiryayev, A. N. (1978), Statistics of Random Processes I, General Theory, Springer, Berlin.

Litterman, R. and Iben, T. (1991), Corporate Bond Valuation and the Term Structure of Credit Spreads, Journal of Portfolio Management, 17, pp.52-64.

Liu, J., Longstaff, F. A., and Mandell, R. E. (2006), The market price of risk in interest rate swaps: The roles of default and liquidity risks, Journal of Business, 79, pp. 2337–2360.

Liu, P. C. (1992), Purchasing power parity in Latin America: A cointegration analysis. Weltwirtschaftliches Archiv, 128, pp. 66-79.

Longstaff, F. A. and Schwartz E. S. (1995a), A simple approach to valuing risky fixed and floating rate debt, The Journal of Finance,.50, pp. 789-819.

Longstaff F A. and Schwartz E S. (1995b), Valuing credit derivatives, The Journal of Fixed Income,.5, pp. 6-12.

Longstaff, A. F., Mithal, S. and Neis, E. (2005), Corporate yield spreads: default risk or liquidity? New evidence from the credit default swap market, The Journal of Finance, 60, pp. 2213-2253.

Lucas, R. E. (1982), Interest rates and currency prices in a two country world, Journal of Monetary Economics, 10, 335-360.

Lucia, J. and Schwartz, E. (2002), Electricity Prices and Power Derivatives: Evidence from Nordic Power Exchange, Review of Derivatives Research, 5, pp. 5-50.

Lund, J. (1997), Econometric analysis of continuous-time arbitrage-free models of the term structure, Working paper, The Aarhus School of Business.

MacKinnon, J. G. (1991), Critical values for cointegration tests, in: Engle, Granger (eds.), pp. 267-276.

Madhur, S. (2002), Costs and benefits of a common currency for ASEAN. ERD Working Paper No. 12, Economic and Research Department, Asian Development Bank.

Malliaris, A. G. and J.L. Urrutia, (1997), Equity and oil markets under external shocks, in: Ghosh, D. and E. Ortiz, eds, Global Structure of Financial Markets, Routledge Publishers, London, U.K., pp.103-116.

Malliaris, A. G. and Urrutia, J. L. (1992), The international crash of October 1987: causality tests, Journal of Financial and Quantitative Analysis, 27, pp. 353-364.

Mallik, G. and Chowdhury, A. (2002), Inflation, Government Expenditure and Real Income in the Long-run, Journal of Economic Studies, 29(2-3), pp. 240-250.

Massey, F. J. (1951), The Kolmogorov-Smirnov Test for Goodness-of-Fit, Journal of the American Statistical Association, 46, pp. 68-78.

McCarthy, J. and Najand, M. (1995), State space modeling of linkages among international markets, Journal of Multinational Financial Management, 5, pp. 1-9.

McDermott, C. J. and Scott, A. (1999), Concordance in business cycles, Reserve Bank of New Zealand Discussion Paper G99/7, Wellington.

McNown, R. and Wallace, M. S. (1989), National price levels, purchasing power parity, and cointegration: A test of four high inflation economies, Journal of International Money and Finance 8, pp. 533-545.

Merton, R. C. (1974), On the pricing of corporate debt: The risk structure of interest rates, Journal of Finance, 29, pp. 449-470.

Merton, R., (1980), On estimating the expected return on the market, Journal of Financial Economics, 8, pp. 323-361.

Mishkin, F. S. (1992), Is the Fisher effect for real? A re-examination of the relationship between inflation and interest rates, Journal of Monetary Economics, 30, pp. 195-215.

Mundell, R. A. (1961), Theory of optimum currency areas. American Economic Review, 51, pp. 663–686.

Musiela, M., Turnbull, S. M. and Wakeman, L.M. (1992), Interest Rate Risk Management, Review of Futures Markets, 12, pp. 221-261.

Nas, T. F and Perry, M. J. (2000), Inflation, inflation uncertainty, and monetary policy in Turkey: 1960-1998, Contemporary Economic Policy, 18 (2), pp. 170-180.

Norden, L. and Weber, M. (2004), Informational efficiency of credit default swap and stock markets: The impact of credit rating announcements. Journal of Banking and Finance, 28, pp. 2813-2843.

Papageorgiou, N. and Skinner, F. S. (2006), Credit Spreads and the Zero-Coupon Treasury Spot Curve, Journal of Financial Research, 29, pp. 421-439.

Pedrosa, M. and Roll, R. (1998), An equilibrium characterization of the term structure, Journal of Fixed Income, 8, pp. 7-26.

Pennacchi G. G. (1991), Identifying the Dynamics of Real Interest Rates and Inflation: Evidence Using Survey Data, The Review of Financial Studies, 4 (1), pp. 53-86.

Perraudin, W., and Taylor, A. (2003), Liquidity and Bond Market Spreads, Working paper, Bank of England.

Phillips, P. C. B. and Mcfarland, J. W. (1997), Forward exchange market unbiasedness: the case of Australian dollar since 1984, Journal of International Money and Finance, 16, 885-907.

Quah D. (1996), Empirics for economic growth and convergence, European Economic Review, 40 (6), pp.1353-1375.

Quah, D. (1993), Empirical cross-section dynamics in economic growth, European Economic Review, 37, pp. 426–434.

Ritchken, P. and Sankarasubramanian, L. (1995), Volatility Structures of Forward Rates and the Dynamics of the Term Structure, Journal of Mathematical Finance, 5, pp. 55-72.

Robert, A. J. and Stuart, M. T. (2000), The intersection of market and credit risk, Journal of Banking & Finance, 24, pp.271-299.

Robert, A. J., David, L. et al. (1997), A Markov model for the term structure of credit risk spreads, The Review of Financial Studies (1986-1998), 10(2), pp. 481-523.

Roberto, B., Simon, B. and Ian, W. M. (2004), An empirical analysis of the dynamic relationship between investment-grade bonds and credit default swaps, Bank of England, Quarterly Bulletin, 44, pp.59-60.

Ruiz, E. (1994), Quasi-maximum likelihood estimation of stochastic volatility models, Journal of Econometrics, 63, pp. 289-306.

Sandmann, G., Koopman, S. J., (1998), Estimation of stochastic volatility models via Monte Carlo maximum likelihood, Journal of Econometrics, 87, pp. 271-301.

Schwartz, E. and Smith, J. E. (2000), Short-term variations and long-term dynamics in commodity prices, Management Science, 46 (7), pp. 893-911.

Silvapulle1, P and Granger, C. W. J. (2001), Large returns, conditional correlation and portfolio diversification: a value-at-risk approach, Quantitative Finance, 1, pp. 542-551.

Simon, H. B. and Nowman, K. B. (1999), Kalman filtering of generalized Vasicek term structure models, Journal of Financial and Quantitative Analysis,.34, pp. 115-130.

Skinner S. and Diaz, A. (2003), An empirical study of credit default swaps, The Journal of Fixed Income, 13 (1). pp. 28 -38.

Skinner, F. S. and Townend, T. G. (2002), An empirical analysis of credit default swaps, International Review of Financial Analysis,.11, pp. 297-309.

Smith, D. R. (2002), Markov Switching and Stochastic Volatility Diffusion Models of Short-Term Interest Rates, Journal of Business and Economic Statistics, 20 (2), pp. 183-197.

Solow, R. (1956), A Contribution to the Theory of Economic Growth, Quarterly Journal of Economics, 70, pp. 65–94.

St Aubyn M. (1995), Evaluating tests for convergence of economic series using Monte Carlo methods with an application on real GDPs per head, Unpublished PhD thesis, London Business School.

St. Aubyn, M. (1999), Convergence across industrialized countries (1890-1989): new results using time series methods, Empirical Economics, 24, pp. 23-44.

Stanton, R. (1997), A nonparametric model of term structure dynamics and the market price of interest rate risk, The Journal of Finance,.52 (5), pp. 1973-2002.

Sun, L. C., (2005), Regime Shifts in Interest Rate Volatility, Journal of Empirical Finance, 12, pp. 418-434.

Tang, Y. (2006), Essays on credit risk, United States -- Texas, The University of Texas at Austin.

Tanizaki, H. (1993), Nonlinear Filters, Estimation and Applications, Lecture Notes in Economics and Mathematic Systems, 400. Berlin: Springer-Verlag.

Taylor, S.J. (1986), Modelling Financial Time Series, Chichester, UK: John Wiley and Sons.

Taylor, M. P. and McMahon, P. C. (1988), Long-run purchasing power parity in the 1920s, European Economic Review, 32, pp. 179-197.

Théoret, R., Rostan, P., and El Moussadek, A. (2004), Forecasting the Canadian Interest-Rate Term Structure: Using the Model of Fong and Vasicek, The Extended Kalman Filter and the Bollinger Bands, Working paper 18.

Van der Merwe, R., Doucet, A. Freitas, N. and Wan, E. (2000), The unscented particle filter, Technical Report CUED/F-INFENG/TR 380.

Vasicek, O. (1977), An equilibrium characterization of the term structure, Journal of Financial Economics, 5, pp. 177-188.

Wall, K. D. (1987), Identification theory for time varying coefficient regression models, Journal of Time Series Analysis, 8, pp. 359-371.

Wan, E., and Van der Merwe, R. (2001), The unscented Kalman filter, in S. Haykins (Ed.), Kalman filtering and neural networks, Wiley New York, pp.221–280.

Wells C. (1996), The Kalman Filter in Finance, Advanced Studies in Theoretical and Applied Econometrics, Kluwer Academic Publishers, 32.

Wilcox, J. A. (1983), Why real interest rates were so low in the 1970's? American Economic Review, 78, pp. 44-53.

Wilmott, P., Dewynne, J. and Howison, S. (1993), Option Pricing: Mathematical Models and Computation, Oxford: Oxford Financial Press.

Wilson, B. K. (2006), The link between inflation, inflation uncertainty and output growth: New time series evidence from Japan, Journal of Macroeconomics, 28, pp. 609-620.

Wilson, B. K. and Culver, S. E. (1999), On measuring the response of real GDP growth to changes in inflation volatility, Quarterly Journal of Business and Economics, 38, pp. 3-15.

Wilson, T. C. (1997a), Measuring and managing credit portfolio risk: Part I: Modelling systemic default risk, The Journal of Lending and Credit Risk Management, 79 (11), pp. 61-72.

Wilson, T. C. (1997b), Measuring and managing credit portfolio risk: Part II: Portfolio loss distributions, The Journal of Lending and Credit Risk Management, 79 (12), pp. 67-78.

Wilson, T. C. (1998), Portfolio credit risk. Economic Policy Review, Federal Reserve Bank of New York, 4, pp. 71-82.

Wolff, C. (1987), Forward exchange rates, expected spot rates, and premia: A signal extraction approach, Journal of Finance, 42, pp. 395-407.

Wolff, C. C. P. (2000), Measuring the forward exchange risk premium: multi-country evidence from unobserved component models, Journal of International Financial markets, Institutions and Money, 10, 1-8.

Wu, Y. and Zhang, H. (1997), Forward premiums as unbiased predictors of future currency depreciation: a non-parametric analysis, Journal of International Money and Finance, 16, 609-623.

Young, P. C., Ng, C. N. and Annitage, P. (1989), A systems approach to economic forecasting and seasonal adjustment, International Journal on Computers and Mathematics with Applications, 18, pp. 481 - 501.

Young, P. C., Pedregal, D. J. and Tych, W. (1999), Dynamic harmonic regression, Journal of Forecasting, 18, pp. 369-394.

Young, P.C. (1988), Recursive extrapolation, interpolation and smoothing of non-stationary time series, in: Chen, C.F. (ed.), Identification and System Parameter Estimation. Pergamon Press, Oxford, pp. 33-44.

Young Ho, E.(2004), Structural Models of Corporate Bond Pricing: An Empirical Analysis, The Review of Financial Studies, 17, pp.499-544.

Yuhn, K. H. (1996), Is the Fisher effect robust? Further evidence, Applied Economics Letters, 3, pp. 41-44.

Zhou, C. (1997), A jump-diffusion approach to modeling credit risk and valuing defaultable securities, Working paper, Federal Reserve Board.

Zhou, S. (1997), Purchasing power parity in high-inflation countries: A cointegration analysis of integrated variables with trend breaks, Southern Economic Journal, 64, pp. 450-467.

Zhou, C. and Qing, C. (2000), A state-space model of short and long-horizon stock returns, The Journal of Financial Research, 23, pp. 523-544.

Index

affine models, 162
aggregate savings rate, 189
amount of risk, 48, 62
arbitrage opportunities, 186, 191

Backward Pass Kalman **Filter, 1**
bootstrap resampling, 125, 135, 145,
    147, 148, 150

Chapman-Kolmogorov equation, 4
common component, 27, 29, 30
common jumps, 285, 300
component structure, 184, 188, 199,
    204
concordance, 50, 55, 62
conditional density, 2, 6
Control Engineering, 2
convenience yield, 286, 288
country-specific features, 25
credit default swap, 156
Credit Default Swaps, 184, 185, 186
credit protection, 191
credit rating premium, 184, 210
credit risk, 156, 157, 159, 170, 185,
    186, 189, 191, 194, 197, 198, 199,
    210, 212
credit spread puzzle, 153, 154
credit spreads, 149, 150, 151, 152, 153,
    154, 155, 157, 158, 159, 160, 161,
    163, 164, 165, 166, 167, 169, 170,
    173, 298

default probabilities, 150, 159
default risk, 149, 152, 153, 154, 157

Deterministic Function of Time, 100
deterministic trend, 190
dynamic factor analysis, 27
Dynamic Harmonic Regression, 1, 17,
    19

earnings yield, 48, 70, 71, 72, 74
economic convergence, 48, 63, 64, 65,
    69, 70, 76
elasticity of variance, 259
Electricity Market, 285, 294, 297
electricity prices, 284, 286, 287, 288
equilibrium spread, 188
Equity market risk premium, 48
equity premium puzzle, 153
equivalent martingale measure, 232
equivalent probability measure, 95, 98,
    118
ex-ante risk premium, 48, 70, 71, 72,
    73, 74
expected inflation, 76, 78, 82, 83, 84,
    85
Explanatory Variables, 195
Extended Kalman Filter, 1, 3, 6, 19,
    240
Extracted Components, 205, 207

Filtering with Jumps, 284
finite memory, 2
finite-dimensional Markovian system,
    94, 96, 99, 115
Fisher Hypothesis, 82
Fixed Interval Smoothing, 1, 19
flight to quality, 155

forecast density, 269, 271, 279
foreign exchange market, 20, 48
forward exchange contract, 31
forward exchange rate, 31, 34, 35, 36,
    41, 43, 44
forward interest rates, 94
Forward Pass Kalman Filter, 1
forward rate volatility, 130, 139

global factor model, 26

Heath-Jarrow-Morton, 94, 95, 114
hedging interest rate exposures, 125
high-yield bonds, 151
Hodrick-Prescott filter, 82

illiquidity, 184, 187, 201, 209
implied volatility, 190
implied volatility, 229, 230, 246, 247,
    249, 250
impulse component, 81
impulse response components, 80
inflation risk, 78
inflation targeting, 76, 77, 78, 83, 84,
    85
inflation uncertainty, 76, 77, 78, 81, 83,
    84, 85
Interest rate futures, 125
iterative process, 3
iTraxx, 188, 191, 192, 193, 194, 195,
    196, 197, 201, 205, 206, 207, 208,
    209, 210, 211, 214, 215, 216, 217,
    218, 219, 220, 221, 222, 223, 224,
    226, 227, 228, 229, 246, 247, 248,
    249, 251, 252, 253, 254, 255, 285,
    298, 299, 300, 301, 302, 303, 304,
    305, 308, 309, 310, 312, 314, 318,
    319, 320, 321, 322

Jacobian matrices, 7, 240
jump amplitude, 290, 295, 297
jump probability, 262
jump risks, 284, 285, 287, 294, 297,
    300, 309
Jumps in CDS Indices, 297

Kalman Filter and Poisson Jumps, 291
Kalman filter and smoother, 261, 268
Kalman filter, 49, 52, 73, 81, 82, 94,
    96, 97, 102, 110, 114, 158, 161,
    164, 204, 284, 291, 292, 305, 307
Kalman gain, 5, 6, 12
Kolmogorov backward equation, 112

leverage ratio, 194
Linear Kalman Filter, 22, 24, 33
liquidity premium, 149, 159
liquidity risk factor, 149
loadings, 158, 160, 161, 162, 164, 165,
    166, 167, 168, 170

market price of interest rate risk, 98,
    108, 111, 117
Market Price of Risk, 37, 44, 47
martingale process, 43
mean reverting process, 8
measurement equation, 3, 5, 6, 7, 9, 10,
    12, 18
mixture of normal distribution, 293
Model for Spot Electricity Prices, 288
Monte Carlo Likelihood, 257, 258
Multifactor Model, 149, 160
Multivariate Analysis of Jumps, 305

nominal uncertainty, 77
non-Markovian, 94, 95, 96, 97, 99, 101,
    104, 105, 111, 114

output gap, 79, 82, 84, 85
output growth, 77

Particle Filter, 3, 13, 15, 19
persistent component, 185, 199, 201,
    202, 203, 205, 206, 208, 209, 210,
    212
pivotal quantity, 137, 147, 150
posterior distribution, 3
preference free, 97, 108, 111
price of risk, 48, 49, 50, 51, 53, 54, 55,
    58, 59, 62, 69, 76
principal components, 152

probability integral transform, 269

Quasi Maximum Likelihood, 257

real exchange rates, 20, 21, 23, 25, 30,
    32, 48
real interest rates, 30, 32
realized volatilities, 249, 250
recovery rate, 189, 190
reduced-form models, 150
risk management, 185, 186, 191
risk-averse investor, 49
riskless arbitrage, 127

scaling parameters, 11
Sigma Points, 11
Signal Processing, 2
slope of the yield curve, 185, 195, 198,
    199, 205, 209, 210
spikes, 286, 287, 288
stochastic filtering, 1
stochastic unit root, 22, 23, 25
stochastic volatility, 230, 257, 258,
    259, 261, 263, 264, 265, 274
structural breaks, 21
Structural uncertainty, 76, 84
subprime market, 193
swaptions, 235
systematic market risk, 154

Theil's inequality coefficient, 269, 279,
    296
theoretical explanatory variables, 189,
    192, 212
time varying risk premia, 35
transient component, 206, 208, 209,
    210, 211
transition equation, 3, 5
transmission of information, 230
treasury slope, 152, 166, 167
two factor short rate model, 8

unbiasedness hypothesis, 35
unit price of risk, 48, 49, 50, 62
unobserved component model, 49
Unobserved Component, 1, 16, 188
unobserved latent variable, 258
unobserved volatility, 1
Unscented Kalman Filter, 1, 3, 10, 19
    229, 231, 241, 243

volatility function, 94, 95, 98, 100, 113,
    125, 126, 127, 129, 130, 131, 133,
    138, 139
volatility of equity, 199

yield spreads, 149, 151, 153, 154, 155,
    156, 157, 159, 166, 168